# EAI/Springer Innovations in Communication and Computing

**Series Editor**

Imrich Chlamtac, European Alliance for Innovation, Ghent, Belgium

The impact of information technologies is creating a new world yet not fully understood. The extent and speed of economic, life style and social changes already perceived in everyday life is hard to estimate without understanding the technological driving forces behind it. This series presents contributed volumes featuring the latest research and development in the various information engineering technologies that play a key role in this process. The range of topics, focusing primarily on communications and computing engineering include, but are not limited to, wireless networks; mobile communication; design and learning; gaming; interaction; e-health and pervasive healthcare; energy management; smart grids; internet of things; cognitive radio networks; computation; cloud computing; ubiquitous connectivity, and in mode general smart living, smart cities, Internet of Things and more. The series publishes a combination of expanded papers selected from hosted and sponsored European Alliance for Innovation (EAI) conferences that present cutting edge, global research as well as provide new perspectives on traditional related engineering fields. This content, complemented with open calls for contribution of book titles and individual chapters, together maintain Springer's and EAI's high standards of academic excellence. The audience for the books consists of researchers, industry professionals, advanced level students as well as practitioners in related fields of activity include information and communication specialists, security experts, economists, urban planners, doctors, and in general representatives in all those walks of life affected ad contributing to the information revolution.

Indexing: This series is indexed in Scopus, Ei Compendex, and zbMATH.

**About EAI -** EAI is a grassroots member organization initiated through cooperation between businesses, public, private and government organizations to address the global challenges of Europe's future competitiveness and link the European Research community with its counterparts around the globe. EAI reaches out to hundreds of thousands of individual subscribers on all continents and collaborates with an institutional member base including Fortune 500 companies, government organizations, and educational institutions, provide a free research and innovation platform. Through its open free membership model EAI promotes a new research and innovation culture based on collaboration, connectivity and recognition of excellence by community.

Ömer Melih Gül • Paolo Fiorini •
Seifedine Nimer Kadry
Editors

# 7th EAI International Conference on Robotic Sensor Networks

EAI ROSENET 2023

Springer

EAI
RESEARCH MEETS INNOVATION

*Editors*
Ömer Melih Gül
Department of Computer Engineering
Bahçeşehir University
İstanbul, Türkiye

Paolo Fiorini
Department of Computer Science
University of Verona
Verona, Italy

Seifedine Nimer Kadry
Department of Applied Data Science
Noroff College
Kristiansand, Norway

International Conference on Robotics and Networks, https://rosenets.eai-conferences.org/2023/

ISSN 2522-8595 ISSN 2522-8609 (electronic)
EAI/Springer Innovations in Communication and Computing
ISBN 978-3-031-64494-8 ISBN 978-3-031-64495-5 (eBook)
https://doi.org/10.1007/978-3-031-64495-5



This Springer imprint is published by the registered company Springer Nature Switzerland AG
The registered company address is: Gewerbestrasse 11, 6330 Cham, Switzerland

# Preface

We are delighted to introduce the proceedings of the first edition of the 7th European Alliance for Innovation (EAI) International Conference on Robotics and Networks (ROSENET 2023). This conference has brought researchers, developers, and practitioners around the world who are leveraging and developing Artificial Intelligence, Computer Vision, Multimedia, and Robotics.

The technical program of ROSENET 2023 consisted of 13 full papers and 2 short papers in oral presentation sessions at the main conference tracks. The conference tracks were Main Track and Short Track. Aside from the high-quality technical paper presentations, the technical program also featured four keynote speeches and one industrial talk. The four keynote speakers were Dr. Kostas Karpouzis from Panteion University of Social and Political Sciences, Greece; Prof. Dr. Mouloud Adel from Aix-Marseille University, France; Nizamettin Aydın from İstanbul Technical University, Türkiye; and Seifedine Nimer Kadry from Noroff College, Norway. The industrial talk was presented by Mr. Emre Bekler from FANUC, Türkiye.

Coordination with the steering chairs, Imrich Chlamtac and Shuau Li, was essential for the success of the conference. We sincerely appreciate their constant support and guidance. It was also a great pleasure to work with such an excellent organizing committee team for their hard work in organizing and supporting the conference. In particular, the Technical Program Committee, led by our TPC Co-Chairs, Dr. Paolo Fiorini, Dr. Seifedine Nimer Kadry, and Dr. Ahmet Onat who have completed the peer-review process of technical papers and made a high-quality technical program. We would like to Dr. Aydan Müşerref Erkmen, Dr. Burak Kantarcı, and Dr. Yücel Batu Salman. We are also grateful to our Sponsorship and Exhibition Chair Dr. Mehmet Berke Gür and our Local Chair Dr. Ali Hamidoğlu. We are also grateful to Conference Manager, Marica Scevlikova, for her support and all the authors who submitted their papers to the ROSENET 2023 conference.

We strongly believe that the ROSENET 2023 conference provides a good forum for all researchers, developers, and practitioners to discuss all science and technology aspects that are relevant to robotics and networks. We also expect that the future ROSENET conference will be as successful and stimulating, as indicated by the contributions presented in this volume.

İstanbul, Türkiye Ömer Melih Gül
Verona, Italy Paolo Fiorini
Kristiansand, Norway Seifedine Nimer Kadry

# Conference Organization

**Steering Committee**

| | |
|---|---|
| Imrich Chlamtac | University of Trento, Italy |
| Shuai Li | Swansea University, UK |

**Organizing Committee**

| | |
|---|---|
| General Chair | |
| Ömer Melih Gül | Bahcesehir University, İstanbul, Türkiye |
| General Co-Chairs | |
| Yücel Batu Salman | Bahcesehir University, İstanbul, Türkiye |
| Aydan Müşerref Erkmen | Middle East Technical University, Ankara, Türkiye |
| Burak Kantarcı | University of Ottawa, Ottawa, ON, Canada |
| TPC Chair and Co-Chair | |
| Paolo Fiorini | University of Verona, Verona, Italy |
| Seifedine Nimer Kadry | Noroff College, Kristiansand, Norway |
| Sponsorship and Exhibit Chair | |
| Mehmet Berke Gür | Bahcesehir University, İstanbul, Türkiye |
| Local Chair | |
| Ali Hamidoğlu | Bahcesehir University, İstanbul, Türkiye |
| Workshops Chair | |
| Alar Kuusik | Tallinn University of Technology, Tallinn, Estonia |
| Publicity & Social Media Chair | |
| Celia Shahnaz | Bangladesh University of Engineering & Technology, Dhaka, Bangladesh |
| Cağan Yılmaz | Dogus University, İstanbul, Türkiye |
| Publications Chair | |
| Ruben Gonzales Crespo | Universidad Internacional de La Rioja, Logrono, Spain |
| Web Chair | |
| Hakan Uysal | Softtech CO, İstanbul, Türkiye |

(continued)

| Technical Program Committee | |
|---|---|
| Gökhan Koray Gültekin | Ankara Yıldırım Beyazıt University, Ankara, Türkiye |
| Mustafa Mert Ankaralı | Middle East Technical University, Ankara, Türkiye |
| Pınar Sarısaray Bölük | Bahcesehir University, İstanbul, Türkiye |
| Tarkan Aydın | Bahcesehir University, İstanbul, Türkiye |
| Ece Gelal Soyak | Bahcesehir University, İstanbul, Türkiye |
| Görkem Kar | University of Central Missouri, Missouri, USA |
| Katarina Trojachanec Dineva | The University "Ss. Cyril and Methodius", Macedonia |
| Alar Kuusik | Tallinn University of Technology, Tallinn, Estonia |
| Islam Tharwat Abdel Halim | Nile University, Egypt |
| Celia Shahnaz | Bangladesh University of Engineering and Technology, Dhaka, Bangladesh |
| Frank Z. Wang | University of Kent, Kent, UK |
| Kostas Karpouzis Panteion | University of Social and Political Sciences in Athens, Greece |
| Ahmed Hassan Mohamed Yousef | Egypt University of Informatics, Egypt |
| Ali Hamidoğlu | Bahcesehir University, İstanbul, Türkiye |
| Erkut Arican | Bahcesehir University, İstanbul, Türkiye |
| Ercan Atam | Bogazici University, İstanbul, Türkiye |
| Ismail Burak Parlak | Galatasaray University, İstanbul, Türkiye |
| Sinan Yiğit | Bahcesehir University, İstanbul, Türkiye |
| Veerpratap Meena | Malavia National Institute of Technology, Jaipur, India |
| Mustafa Üstüner | Artvin Coruh University, Türkiye |

# Contents

# Part I
# Robotics

# A Novel Nash-Based Matching Approach for Multirobot Task Allocation in Distributed Robotic Networks

Ali Hamidoğlu, Ömer Melih Gül, Gökhan Koray Gültekin, and Seifedine Nimer Kadry

## 1 Introduction

### *1.1 Motivation*

Networks of mobile robots with abundant resources have recently been used in many fields with wireless sensors. Even in dangerous or sparsely populated areas, robots can be used to carry out automated operations without human involvement. Volcanic regions may be one of the applications for static sensor nodes because these regions have ideal conditions for measuring variables like temperature and gas concentration. An emergency scenario called an "event" arises when parameters are detected to have dramatically changed. It is necessary to assess events using a robot fitted with more precise sensing tools, such as seismometers, thermal cameras, etc [1].

A. Hamidoğlu (✉)
Department of Mathematics, Bahcesehir University, İstanbul, Türkiye
e-mail: ali.hamidoglu@eng.bau.edu.tr

Ö. M. Gül
Department of Computer Engineering, Bahçeşehir University, İstanbul, Türkiye
e-mail: omermelih.gul@bau.edu.tr

G. K. Gültekin
Department of Electrical-Electronics Engineering, Ankara Yıldırım Beyazıt University, Ankara, Türkiye
e-mail: gkgultekin@aybu.edu.tr

S. N. Kadry
Department of Applied Data Science, Noroff College, Kristiansand, Norway

Artificial Intelligence Research Center (AIRC), Ajman University, Ajman, UAE
e-mail: seifedine.kadry@noroff.no

 
Ö. M. Gül et al. (eds.), *7th EAI International Conference on Robotic Sensor Networks*, EAI/Springer Innovations in Communication and Computing,
https://doi.org/10.1007/978-3-031-64495-5_1

Researchers have recently looked into deployment of robots in a variety of applications, such as cleaning up dangerous regions, harvesting fields, exploring new places, keeping an eye on the battlefield, performing search and rescue operations, etc. Due to their sensitivity to failure and the length of time required to complete tasks, such complicated domains reduce the likelihood of using a single robot. Multirobot systems are driven by the need to complete difficult missions since they are more fault-tolerant and resilient [2]. There is a need for distributed solutions where numerous robots operate autonomously, break down larger jobs into smaller ones, bargain (and renegotiate in case of failure) for the assignment of smaller tasks, and do the smaller tasks as rapidly as feasible.

Wireless sensors and robots complement one other in numerous ways. Robotic and wireless sensor network (RWSN) is the name given to their combination, which has a variety of applications in the military, transportation monitoring, healthcare, weather forecasting, autonomous driving, mining, and search and rescue, to name just a few [3].

For multirobot systems to function autonomously, the multirobot task allocation (MRTA) problem must be effectively solved. The fundamental solution to this problem, commonly referred to as the "linear assignment problem," is as follows: *Given a set of agents (or robots) and tasks, with each agent obtaining some payoff (or incurring some cost) for each task, find a one-to-one assignment of agents to tasks so that the overall payoff of all the agents is maximized (or cost is minimized)* [4].

Numerous fields, such as autonomous exploration [5], logistics [6], unmanned search and rescue [7], reconnaissance [8], etc., may find use for the MRTA problem whose solution is generally known to be NP-hard [9]. As a result, it should be remembered that the quality of proposed solutions is inversely related to their temporal complexity. Our study is motivated by a case study from [10] that shows how multirobot systems, industrial applications, and workload distribution are related.

As described in [11–13], multirobot systems that use RF-domain security solutions may require computational and hence energy resources if they employ deep learning techniques. As a result, task execution in multirobot systems requires more energy-efficient methods.

To increase network lifetime of a multirobot system, energy efficiency is considered in the work [14]. The paper considers the traveled distance and battery level of all robots and the predicted harvested energy of all robots.

In the discipline of engineering and management science, game theory, a branch of mathematics and economics, has emerged as a valuable tool. Game theory provides a framework for analyzing strategic interactions among decision-makers, including individuals, businesses, and governments. By simulating these interactions as games, managers can obtain insight into decision-making processes, predict the behavior of rivals, and optimize their own strategies. Game theory has numerous applications in the field of engineering and management science [15, 16], including pricing strategies [17, 18], negotiation tactics [19], supply chain management [20], and military tactics [21]. In Nash-based games, players make informed decisions by considering not only their own actions but also the potential reactions of others. This strategic perspective enables organizations to make better decisions in complex and competitive environments, leading to improved outcomes, enhanced efficiency, and increased profits. This paper explores a novel Nash-based approach that employs game theory to solve the MRTA problem, in which robots cooperate to divide up duties. In this regard, robots establish equilibrium decision-making on the MRTA problem, resulting in a low computational cost in comparison to the Hungarian algorithm [22], which is regarded as a benchmark.

### *1.2 Our Contributions*

Main contributions of this paper can be summarized as follows:

- A formulation of the energy consumption of a robot in motion is investigated in conjunction with the optimal speed for maximizing the amount of energy harvested by a robot while minimizing its total motion energy during the MRTA process.
- A novel cooperative Nash game based on distance is proposed in order to plan a near-optimal matching solution for MRTA problems.
- Using the Nash equilibrium strategy, a novel distributed matching algorithm is developed to provide a cost-effective solution for the MRTA problem.

### *1.3 Organization*

The rest of this paper is structured as follows. In Sect. 2, we present the system model. Section 3 establishes an energy consumption model together with harvested energy where the optimal speed of a robot is derived. Section 4 provides a novel cooperative Nash game in the context of MRTA, accompanied by the key findings. In Sect. 5, the suggested model is tested and compared with the Hungarian algorithm on various MRTA scenarios including 10 and 20 robots, each with the same number of jobs. The paper is concluded in Sect. 6, which also suggests areas for additional investigation.

## 2 System Model

The MRTA problem in distributed robotic networks (DRN) is addressed in this research. The topic of MRTA in DRNs entails the effective allocation of a collection of autonomous robots to a given set of tasks, with the aim of optimizing specific

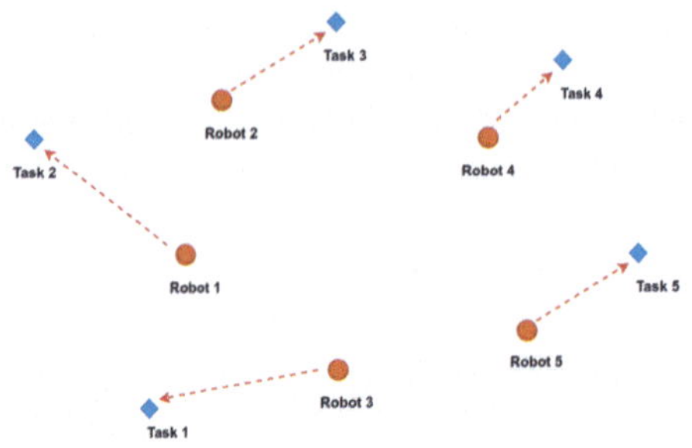


**Fig. 1** A networked system with five robots is an example system where each robot chooses a task to complete in each round. The dashed lines show the matchings of robot-task pairs. For example, Task 2 is assigned to Robot 1

performance parameters. In the following sections, we provide the system model and problem description regarding the MRTA problem.

Typically, a system model for multirobot task allocation through a matching algorithm involves defining the system's components and processes. Here is a brief description of the system:

- *Robots*: A set of autonomous robots, denoted as $R = R1, R2, \ldots, Rn$, each equipped with sensors, actuators, and communication capabilities.
- *Tasks*: A set of tasks that are distributed among robots, denoted as $T = T1, T2, \ldots, Tn$.
- *Environment*: The physical or virtual platform on which robots and tasks are located. Here, it is assumed to be a two-dimensional plane.

In our context, we consider the following system processes in our study:

- *Task allocation:* Based on the compatibility of robot skills and task requirements, the matching algorithm assigns robots to tasks. In this framework, we seek to minimize the total distance between assignments and robots to reduce completion time.
- *Matching algorithm*: By taking into account robot capabilities and task requirements as inputs and generating an assignment matrix, the matching algorithm determines which robot is assigned to which task. In this particular context, we present a novel matching algorithm based on Nash equilibrium alongside the Hungarian algorithm, which serves as a benchmark due to its status as the optimal matching algorithm for the MRTA problem (Fig. 1).

## 3 Formulation of Energy in Task Allocation

### *3.1 Energy Consumption*

Regarding the motion system, the energy consumption can be primarily broken down into three parts: kinetic energy, friction energy, and thermal energy. There are three stages to a robot's motion: standby, startup, and sustained operation. In standby mode, the power supply is constant. An instantaneous pulse is required to transmit the start signal to the electric motors during the startup phase. When a robot begins to move, it enters secure operation. More precisely, overall energy consumption is formulated as follows [23]:

$$E_{consumption} = E_{kinetic} + E_{friction} + E_{termal} \tag{1}$$

where

$$E_{kinetic} = \frac{1}{2}Mv^2; \tag{2}$$

$$E_{friction} = \int \eta M v dt; \tag{3}$$

$$E_{termal} = \int (\sigma_1 t^2 + \sigma_2 t + \sigma_3 v) dt. \tag{4}$$

In this context, the variable $M$ denotes the weight of the robot, while $v$ represents its speed. The symbol $\eta$ is used to denote the coefficient of friction between the robot's wheel and the ground. Additionally, the variables $\sigma_1$ and $\sigma_2$ are employed to represent coefficients that relate to the relationship between time and heat. Lastly, the variable $\sigma_3$ is used to indicate the coefficient that corresponds to the link between speed and heat in the framework of the robot.

## 3.2 *Energy Harvesting*

In the present context, we investigate a robot characterized by a consistent speed of motion that acquires energy along its trajectory. Thus, the process of energy harvesting is directly influenced by the speed and mass of the robot. In this study, we examine a specific energy harvesting formulation for a robot involved in the work allocation process:

$$E_{harvest} = \int \mu M v dt \tag{5}$$

where $\mu$ represents the proportionality constant that is influenced by both the speed and mass of the robot during the motion.

## 3.3 *Optimal Speed*

In this section, we explore the concept of achieving an appropriate speed profile for a robot's movement, with the objective of maximizing energy harvesting while simultaneously reducing energy consumption. More precisely, we consider the following problem:

**Problem 1 (Optimal Speed)** *Determine the optimal speed of a robotic system so as to maximize energy harvest while minimizing energy consumption:*

$$\max_{v\in\mathbb{R}^+}\Big(\int \mu M v dt - \frac{1}{2}Mv^2 - \int \eta M v dt - \int (\sigma_1 t^2 + \sigma_2 t + \sigma_3 v) dt\Big) \tag{6}$$

**Theorem 1** *The optimal speed is*

$$v = \int \Big(\mu - \eta - \frac{\sigma_3}{M}\Big) dt, \tag{7}$$

*for the Problem 1.*

## 4 A Cooperative Nash Game

In this part, we present a novel cooperative Nash game formulation for the MRTA problem. This game assumes that robots function as players, with each player aiming to achieve an equilibrium approach for the allocation of duties within the group. Listed below are the game's playability conditions:

1. Each player and task are located on the plane.
2. Locations of $i$th player and task are given as $R_i = (x_i, y_i)$ and $T_i = (w_i, z_i)$, respectively.
3. Each player shares the optimal speed $v^*$ based on the Problem 1.
4. Each player takes only one task by reaching its position.
5. Equilibrium position $E = (e_1, e_2)$ is determined according to the shortest distance from each participant as a group.
6. Each participant collaborates to reduce the distance from the equilibrium point to themselves, which is subtracted from the distance from the task location to the equilibrium point.

In the context of this collaborative game, all participants engage in joint planning to ascertain an equilibrium position, which refers to an optimal location in terms of distance from each individual player. In addition, we establish the set $I_n$ as the collection of $n$ points $(i, j)$ randomly selected from the following permutation set:

$$P_n := \{(i, j) : i, j = 1, 2, \cdots, n\}. \tag{8}$$

Therefore, the set $I_n$ has a cardinality of $n$ points, while the set $P_n$ comprises a cardinality of $n^2$ points. Before stating the problem, we define the concept of paring as

$$\rho_i^j := (R_i, T_j) \quad \text{for} \quad i, j \in \{1, 2, \cdots, n\}, \tag{9}$$

which means the $i$th robot picks the $j$th task in the game. In addition, we define an arbitrary allocation strategy in the game, which consists of $n$ parings as follows:

$$S_i^j = (\rho_{i_1}^{j_1}, \rho_{i_2}^{j_2}, \cdots, \rho_{i_n}^{j_n}), \quad \text{for} \quad i_k, j_k, k \in \{1, 2, \cdots, n\}.$$

Now, we propose the following problem in the context of the cooperative Nash game:

**Problem 2 (Cooperative Nash Policy)** *Minimize the following distance of tasks and robots from the equilibrium point*

$$\min_{S_i^j} \sum_{k=1}^{n} \Big| ||R_{i_k} - E|| - ||T_{j_k} - E|| \Big| \tag{10}$$

*under the playability conditions from 1 to 6.*

**Theorem 2 (An Equilibrium Location)** *In this game, there exists a unique equilibrium position $E = (e_1, e_2)$ given as*

$$e_1 = \frac{1}{n}\sum_{i=1}^{n} x_i \quad \text{and} \quad e_2 = \frac{1}{n}\sum_{i=1}^{n} y_i \tag{11}$$

*which is the shortest distance from each participant as a group.*

The following result is the principal finding of the study, which demonstrates the existence of Nash equilibrium for the Problem 2.

**Theorem 3 (Nash Equilibrium)** *Let $E$ be the equilibrium point as described in* (11). *Assume without loss of generality that*

$$||R_1 - E|| \geq ||R_2 - E|| \geq \cdots \geq ||R_n - E||$$
$$||T_1 - E|| \geq ||T_2 - E|| \geq \cdots \geq ||T_n - E||,$$

*Then, there exists a Nash equilibrium $S_i^i = (\rho_1^1, \rho_2^2, \cdots, \rho_n^n)$ which solves the Problem 2.*

Here, we provide the following result which asserts that the near-optimal Nash policy becomes the optimal in the particular case.

**Theorem 4 (Optimal Nash Policy)** *In the cooperative Nash game, the equality of the following near-optimal Nash policy $S_\ell^\ell$ holds*

$$\min_{S_i^j} \sum_{k=1}^{n} \sqrt{(x_{i_k} - w_{i_k})^2 + (y_{j_k} - z_{j_k})^2} \leq \sum_{k=1}^{n} \sqrt{(x_{\ell_k} - w_{\ell_k})^2 + (y_{\ell_k} - z_{\ell_k})^2}, \tag{12}$$

*if each task equidistantly rests in between equilibrium point and each robot, i.e.,*

$$T_{\ell_i} = \frac{R_{\ell_i} + E}{2} \quad \textit{for all} \ \ i \in \{1, 2, \cdots, n\}. \tag{13}$$

## *4.1 Computational Complexity*

This subsection investigates the computational complexity of the proposed game-theoretical approach as in the following proposition.

**Proposition 1** *The computational complexity of the proposed Nash-based approach is $O(N)$.*

***Proof*** In the proposed Nash-based approach, first a Nash equilibrium point is found by considering the locations of all robots, which brings $N$ computations. Then, the distance of this equilibrium point to location of each task is calculated, which brings $N$ computations. Then, we made $N$ summations, which brings $N$ computations. Finally, $N$ number of distances are sorted twice with a complexity of $2 * O(\log(N))$. To sum up, $3 * N + 2 * O(\log(N)) = O(N)$ computations are required.

Hence, computational complexity of the proposed Nash-based approach is $O(N)$.

As a centralized MRTA approach, the Hungarian algorithm has a computational complexity of $O(N^3)$. On the other hand, as distributed MRTA approach, the proposed game-theoretical approach in this paper has a linear computational complexity of $O(N)$. Hence, our distributed MRTA approach is considerably better than the Hungarian MRTA approach in terms of the computational complexity.

# 5 Numerical Results

In this part, we present two distinct scenarios whereby 10 and 20 robots are allocated to 10 and 20 assignments, respectively, in accordance with the proposed Nash policy. In each scenario, the result is compared with the Hungarian algorithm [22], which is the ground-breaking solution for the MRTA problem, as indicated in Sect. 2. Here, letters, R, T, and E are used to describe the position of a robot, a task, and an equilibrium point, respectively (Table 1).

**Table 1** Sample parameter values for the robot

| Parameter | Value |
|---|---|
| $M$ | 6.5 |
| $\mu$ | 1 |
| $\eta$ | 0.5 |
| $\sigma_3$ | 0.001 |
| $T_0$ | 0 |
| $T_f$ | 10 |

For determining the optimal speed, we use the values given in the following sample parameter table [23]:

By using (7), we calculate the optimal speed as

$$v = \int_0^{10} \left(1 - 0.5 - \frac{0.001}{6.5}\right) dt = 0.4998. \tag{14}$$

In this study, we employ this optimal speed for all three cases as a means to quantify the duration of the task allocation process in separate ways.

## 5.1 First Scenario: 10 Robots vs. 10 Tasks

In this part, we consider ten robots against ten tasks randomly distributed in a 1000 m × 1000 m field. In this regard, the locations of ten robots and tasks are given as

$$\begin{aligned}
&((x_{R1}, y_{R1}), (x_{R2}, y_{R2}), (x_{R3}, y_{R3}), (x_{R4}, y_{R4}), (x_{R5}, y_{R5}), (x_{R6}, y_{R6}), (x_{R7}, y_{R7}),\\
&(x_{R8}, y_{R8}), (x_{R9}, y_{R9}), (x_{R10}, y_{R10})) = ((-318, -236), (-370, -400), (370, 80),\\
&(150, -355), (253, 200), (-200, 230), (-98, -424), (-260, -377),\\
&(-116, -200), (-83, -80)).((x_{T1}, y_{T1}), (x_{T2}, y_{T2}), (x_{T3}, y_{T3}), (x_{T4}, y_{T4}),\\
&(x_{T5}, y_{T5}), (x_{T6}, y_{T6}), (x_{T7}, y_{T7}), (x_{T8}, y_{T8}), (x_{T9}, y_{T9}), (x_{T10}, y_{T10}))\\
&= ((340, 399), (30, -70), (300, 0), (250, 150), (0, 110), (-189, 424),\\
&(250, -100), (442, 457), (75, -441), (-265, -147)).
\end{aligned}$$

A Nash strategy is shown as a near-optimal policy of the MRTA in Fig. 2. Table 2 reveals that the Nash policy yields a route that is 78 meters longer compared to the route chosen by the Hungarian policy. Consequently, this leads to an additional 15 seconds of travel time for the MRTA. On the other hand, the Nash strategy offers a computationally efficient solution in terms of order, as compared to the benchmark.

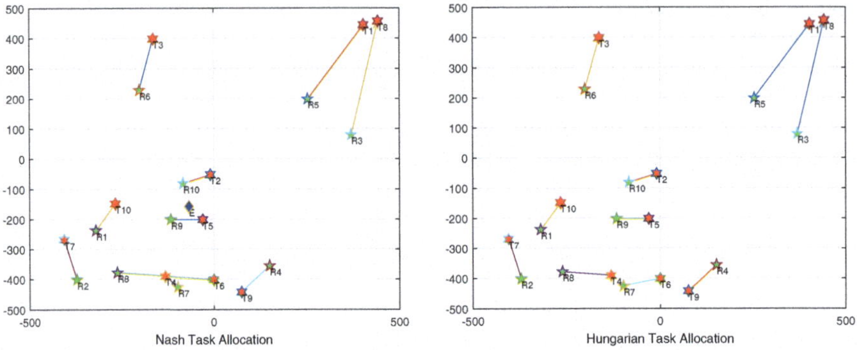


**Fig. 2** The Hungarian strategy provides the optimal policy, while the Nash-based matching algorithm establishes a near-optimal policy in the 10-robot scenario

**Table 2** The performances of the Nash-based approach and the Hungarian approach in the 10-robot scenario

| Policy | Total distance (m) | Total time (sec) | Speed (m/sec) |
|---|---|---|---|
| Nash | 1673 | 335 | 4.998 |
| Hungarian | 1595 | 320 | 4.998 |

## 5.2 Second Scenario: 20 Robots vs. 20 Tasks

In this scenario, the locations of twenty robots and tasks are given as

$$((x_{R1}, y_{R1}), (x_{R2}, y_{R2}), (x_{R3}, y_{R3}), (x_{R4}, y_{R4}), (x_{R5}, y_{R5}), (x_{R6}, y_{R6}), (x_{R7}, y_{R7}), (x_{R8}, y_{R8}), (x_{R9}, y_{R9}), (x_{R10}, y_{R10}), (x_{R11}, y_{R11}), (x_{R12}, y_{R12}), (x_{R13}, y_{R13}), (x_{R14}, y_{R14}), (x_{R15}, y_{R15}), (x_{R16}, y_{R16}), (x_{R17}, y_{R17}), (x_{R18}, y_{R18}), (x_{R19}, y_{R19}), (x_{R20}, y_{R20})) = ((-50, 50), (-123, 75), (-15, 10), (-280, 180), (303, -220), (-353, 130), (410, -50), (-435, 100), (-59, -450), (455, 350), (-233, -170), (200, 200), (123, 232), (-220, -271), (-11, -111), (-79, -156), (85, 185), (190, 100), (225, 50), (-133, -234)).((x_{T1}, y_{T1}), (x_{T2}, y_{T2}), (x_{T3}, y_{T3}), (x_{T4}, y_{T4}), (x_{T5}, y_{T5}), (x_{T6}, y_{T6}), (x_{T7}, y_{T7}), (x_{T8}, y_{T8}), (x_{T9}, y_{T9}), (x_{T10}, y_{T10}), (x_{T11}, y_{T11}), (x_{T12}, y_{T12}), (x_{T13}, y_{T13}), (x_{T14}, y_{T14}), (x_{T15}, y_{T15}), (x_{T16}, y_{T16}), (x_{T17}, y_{T17}), (x_{T18}, y_{T18}), (x_{T19}, y_{T19}), (x_{T20}, y_{T20})) = ((25, 25), (-66.5, 37.5), (-7.5, 5), (-140, 90), (156.5, -110), (-176.5, 65), (205, -25), (217.5, 50), (-29.5, -225), (227.5, 175), (-116.5, -85), (100, 100), (-61.5, 116), (-110, -135.5),$$

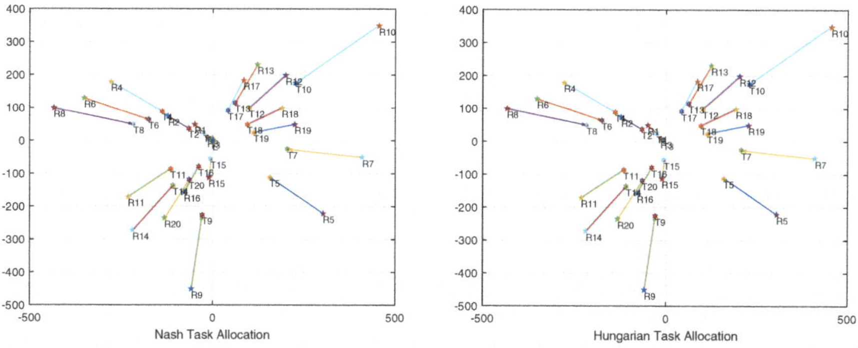


**Fig. 3** The Hungarian strategy matches with Nash-based scheduling in the 20-robot scenario

**Table 3** The performances of the Nash-based approach and the Hungarian approach in the 20-robot scenario

| Policy | Total distance (m) | Total time (sec) | Speed (m/sec) |
|---|---|---|---|
| Nash | 2787 | 558 | 4.998 |
| Hungarian | 2787 | 558 | 4.998 |

$$(-5.5, -55.5), (-39.5, -78), (42.5, 92.5), (95, 50), (112.5, 25), (-65.5, -117)).$$

In this case, the locations of robots and tasks are designed according to the relation (13) in Theorem 4.

Based on the observations presented in Fig. 3, we numerically see that the Nash equilibrium strategy and the benchmark approach align in terms of achieving the optimum tracking scenario. This match serves as a justification for the near-optimal policy of the proposed model, as demonstrated in Theorem 4.

The efficacy of the Nash policy in terms of computational complexity is seen in Table 3. As previously mentioned, the computational complexity of the Nash policy in the MRTA process, which involves a system of $N$ robots and tasks, is of the order $N$. Compared to that, the benchmark complexity for this process is of the order $N^3$.

## 6 Conclusion

The paper introduces a new cooperative Nash game framework that functions as a distributed matching approach for tackling the task allocation problem in a distributed robotic network comprising of robots and tasks. The existence of Nash equilibrium provides a potential cooperation between robots to minimize the overall distance for the assignment problem. Numerical results demonstrate the reliability of the theoretical findings and the robustness of the proposed model in terms of the order of computation.

In the future, the proposed game model will be extended to the solutions for the MRTA problems where robots will cooperatively work together to not only minimize the total distance but also utilize the overall battery efficiently.

## References


1. Lukic, M., Barnawi, A., Stojmenovic, I.: Robot coordination for energy-balanced matching and sequence dispatch of robots to events. IEEE Trans. Comput. **64**(2), 1416–1428 (2015)
2. Gautam, A., Thakur, A., Dhanania, G., Mohan, S.: A distributed algorithm for balanced multi-robot task allocation. In: 2016 11th International Conference on Industrial and Information Systems (ICIIS), pp. 622–627, Roorkee, India (2016). https://doi.org/10.1109/ICIINFS.2016.8263014
3. Ryu, J.H., Irfan, M., Reyaz, A.: A review on sensor network issues and robotics. J. Sens. **2015**, 1–14 (2015).
4. Luo, L., Chakraborty, N., Sycara, K.: Provably-good distributed algorithm for constrained multi-robot task assignment for grouped tasks. IEEE Trans. Robot. **31**(1), 19–30 (2015)
5. Dasgupta, P.: Multi-robot task allocation for performing cooperative foraging tasks in an initially unknown environment. In: Jain, L.C., Aidman, E.V., Abeynayake, C. (eds.) Innovations in Defence Support Systems-2: Socio-Technical Systems, pp. 5–20. Springer, Berlin, Germany (2011)
6. Lim, S., Rus, D.: Stochastic motion planning with path constraints and application to optimal agent, resource, and route planning. In: Proc. IEEE Int. Conf. Robot. Autom., pp. 4814–4821 (2012)
7. Jones, E.G., Dias, M.B., Stentz, A.: Time-extended multi-robot coordination for domains with intra-path constraints. Auton. Robot. **30**(1), 41–56 (2011)
8. Lenagh, W., Dasgupta, P., Munoz-Melendez, A.: A spatial queuing based algorithm for multi-robot task allocation. Robotics **4**(3), 316–340 (2015)
9. Gerkey, B.P., Mataric, M.J.: A formal analysis and taxonomy of task allocation in multi-robot systems. Int. J. Robot. Res. **23**(9), 939–954 (2004)
10. Zitouni, F., Harous, S., Maamri, R.: A distributed approach to the multi-robot task allocation problem using the consensus-based bundle algorithm and ant colony system. IEEE Access **8**, 27479–27494 (2020)
11. Comert, C., Kulhandjian, M., Gul, O.M., Touazi, A., Ellement, C., Kantarci, B., D'Amours, C.: Analysis of Augmentation Methods for RF Fingerprinting under Impaired Channels. In: Proceedings of the 2022 ACM Workshop on Wireless Security and Machine Learning, pp. 3–8, San Antonio, USA (2022)
12. Gul, O.M., Kulhandjian, M., Kantarci, B., Touazi, A., Ellement, C., D'Amours, C.: On the Impact of CDL and TDL Augmentation for RF Fingerprinting under Impaired Channels. In: Proceedings of the 48th Wireless World Research Forum, pp. 1–6, Abu Dhabi, UAE (2022)
13. Gul, O.M., Kulhandjian, M., Kantarci, B., Touazi, A., Ellement, C., D'Amours, C.: Fine-grained Augmentation for RF Fingerprinting under Impaired Channels. In: Proceedings of the IEEE 27th International Workshop on Computer Aided Modeling and Design of Communication Links and Networks, pp. 1–6, Paris, France (2022)
14. Gul, O. M.: Energy harvesting and task-aware multi-robot task allocation in robotic wireless sensor networks, Sensors **23**(6), 3284 (2023)
15. Hamidoğlu, A., Taghiyev, M.H.: On discrete game models with applications to management. In: Proceedings of the Fifteenth International Conference on Management Science and Engineering Management, vol. 1/15, pp. 195–207 (2021)
16. Hamidoğlu, A., Taghiyev, M.H., Weber, G.W.: On building two-player games with treatment schedules for the SIR model. Azerbaijan J. Math. **11**(2), 183–195 (2021)

17. Hamidoğlu, A.: A game theoretical approach for finding near-optimal solutions of an optimization problem. Optimization **72**(10), 2561–2583 (2022)
18. Hamidoğlu, A.: A novel one target game model in the life insurance market. Int. J. Manag. Sci. Eng. Manag. **16**(3), 221–228 (2021)
19. Hamidoğlu, A. Taghiyev, M.H., Weber, G.W.: On construction of pursuit-evasion games in discrete control models. Appl. Comput. Math. **21**(1), 52–60 (2022)
20. Hamidoğlu, A.: A game-theoretical approach on the construction of a novel agri-food supply chain model supported by the government. Expert Syst. Appl. **237**, 121353 (2024)
21. Hamidoğlu, A.: Designing discrete-time control-based strategies for pursuit-evasion games on the plane. Optimization, 1–30 (2023)
22. Kuhn, H.W.: The hungarian method for the assignment problem. Nav. Res. Logist. Q. **2**, 83–97 (1955)
23. Hou, L., Zhang, L., Kim, J.: Energy modeling and power measurement for mobile robots. Energies **12**(1), 27 (2018)

# Learning Soft Robotic Arm Control: A Data-Driven Approach with Forward Dynamics Transformer and Reinforcement Learning

Abdelrahman Alkhodary and Berke Gur

## 1 Introduction

Soft robotic arms exhibit unique properties that make them suitable for a wide range of applications, from minimally invasive surgery to adaptable manipulation in unstructured environments. The ability to control these soft robotic arms effectively is paramount for their successful deployment. In the context of soft robotic arm control, significant progress has been achieved by harnessing supervised learning algorithms. Holst et al. focused on solving the challenge of reaching specific points in space by learning the inverse kinematic model through polynomial regression. Their approach utilized the arm's shape as input and the corresponding cable displacement as output [1]. Furthermore, Bern and colleagues [2] made their contribution by initially establishing the forward kinematic model of the soft robot arm based on empirical data. Subsequently, they employed an optimization algorithm to deduce the inverse model. Employing a gradient-based optimization approach, they successfully determined the desired end-effector position by minimizing the discrepancy between the output of the inverse kinematic neural networks and the target position corresponding to the control inputs. Gillespie et al. used a neural network architecture to enhance the control of soft robotic arms [3]. They utilized the control signal and the current end-effector position as inputs while the outputs

This research was supported by The Scientific and Technological Research Council of Türkiye (TÜBİTAK) through the project entitled "Robotic Squid for Underwater Manipulation and Intervention" (grant no. 216M201). The funder had no role in study design, data collection and analysis, decision to publish, or preparation of the manuscript.

A. Alkhodary (✉) · B. Gur
Bahcesehir University, İstanbul, Türkiye
e-mail: abdelrahman.alkhoda@bahcesehir.edu.tr; berke.gur@bau.edu.tr



Ö. M. Gül et al. (eds.), *7th EAI International Conference on Robotic Sensor Networks*, EAI/Springer Innovations in Communication and Computing,
https://doi.org/10.1007/978-3-031-64495-5_2

were the same information projected into the subsequent time step. This approach allowed the creation of a state matrix and an input matrix for model predictive control, resulting in precise control of the soft robot arm. The authors of [4] expanded on neural network-based control by developing an inverse kinematic algorithm designed for soft robot arms. They established a direct relationship between the pressure applied to rubber hoses and the arm's tip position, enabling effective control signal determination for end-effector relocation.

It's important to note that inverse kinematics, as shown by Holsten et al. [1] and Kovandzic et al. [4], often require complex models that can accurately represent the soft arm systems. In contrast, certain reinforcement learning algorithms offer a model-free, data-driven alternative, allowing autonomous learning of optimal control policies for robotic arms through trial and error. This research introduces a data-driven approach utilizing the forward dynamics transformer (DynaFormer) and reinforcement learning to perform soft robotic arm control. Reinforcement learning (RL) algorithms have been used to learn the control of soft robotic arms. In their work, Ansari et al. introduced an innovative assistive manipulator using RL, designed to automate bathing for the elderly [5]. In this research, the soft arm consisted of one proximal module and two distal modules. The distal module incorporated three pneumatic actuators and tendons, and their control strategy used cooperative multi-agent reinforcement learning, with each actuator functioning as an autonomous agent. Similarly, in [6] researchers applied RL to control a Honeycomb PneuNets soft robot. The authors trained the RL model in a simulated environment and transferred the knowledge to the physical robot. Satheeshbabu et al. [7] extended this to continuum arm control, using deep Q-learning (DQN) [8] for open-loop position control. In [9], Satheeshbabu et al. explored model-free reinforcement learning with deep deterministic policy gradients (DDPG) [10], enhancing soft manipulator control for larger state-action spaces. A different research group, Centurelli and colleagues also developed a neural network-based closed-loop controller for soft manipulators using reinforcement learning [11]. The controller is trained in simulation using a long-short term memory (LSTM) network to approximate the robot's forward dynamic model. The results demonstrate the effectiveness of the controller in normal working conditions and with a payload attached to the manipulator's end effector. Reinforcement learning agents naturally operate in a sequential manner, where an agent observes the state of the environment, takes an action accordingly, and then receives a reward based on the action. Then, the environment transits to the next state. In that sense, Chen et al. formulated offline reinforcement learning as a conditional sequence model, named the decision transformer [12]. The decision transformer is an autoregressive model that generates the action predictions based on the past experience of an agent composed of sequence of state-action-rewards by utilizing a causally masked transformer. This transformer is constructed based on the GPT transformer framework [13], which is in turn rooted in the decoder component of the original transformer architecture as introduced by Vaswani et al. [14]. For online reinforcement learning paradigm, the authors of [15] proposed Deep Transformer Q-Network (DTQN), an architecture that leverages the transformer architecture and self-attention to extend the DQN [8] algorithm to solve partially observable reinforcement learning domains.

This work presents a framework to learn a control policy for soft robotic arm directly from data collected from the arm. The framework starts first by learning the forward dynamics of the soft robotic arm with data collected from the real arm. This approach represents one of the main contributions of this research. The forward dynamics model (dubbed the dynamic transformer or DynaFormer) is learned using a transformer-based architecture. Using the learned forward dynamics (or DynaFormer), a RL agent is learned using twin-delayed deep deterministic policy gradient algorithm (TD3) [16]. The task for the RL agent is to reach a desired point. The process will start by collecting data from soft robotic arm from a simulation environment, followed by training the forward dynamic model on this collected data. The trained forward model acts as the simulation environment for the RL algorithm. The remainder of the paper is structured as follows. The proposed forward dynamic model is explained in Sect. 2 followed by Sect. 3 where the learning framework employing the RL TD3 algorithm is described. Section 4 presents the experiments conducted to evaluate the learning framework. Lastly, concluding remarks for the paper are presented in Sect. 5.

## 2 Forward Dynamics Model

Modeling the forward dynamics of a robotic arm is of significant importance in various applications, including arm control and simulation. This paper introduces DynaFormer, a novel model designed with the aim of learning the forward dynamics of soft robotic arm. The DynaFormer model is a data-driven technique that utilizes the transformer architecture. It is a generative model that is founded on the decoder component of the original transformer architecture proposed by Vaswani et al. [14]. The outputs of the DynaFormer are produced in accordance with the preceding inputs. The input to the DynaFormer model is a sequence of $N$-element list $\Sigma = [\tau_0, \tau_1, \cdots, \tau_{N-1}]$ with each element defined as $\tau_n = [S_n, A_n, P_n]$, where $S_n$ is the state of the arm, $A_n$ is the action, and the $P_n$ is the position of the tip of the arm at time step $n$ for $0 \leq n \leq N-1$ (see Fig. 1). The time steps are assumed to be constant. The current state of the arm $S_n = [A_{n-1}, P_{n-1}]$ consists of the action $A_{n-1}$ performed on the last time step $n-1$ and the tip position result $P_{n-1}$ from that action. Similarly, $A_n$ is the action performed at time step $n$, whereas $P_n$ is the tip position of the soft robotic arm that results from the action $A_n$. This tip position $P_n$ is masked at time step $n$ so that it can be predicted. The state of the arm $S_n$ at time step $n$ can be assumed as auxiliary element as it consists of the action and the resulting position of the tip of the arm at time step $n-1$. However, it was found that $S_n$ improves the prediction of tip position the arm. Therefore, $S_n$ is also included as a part in the element $\tau_n$. The internal structure of the DynaFormer model is explained in detail in the following section.

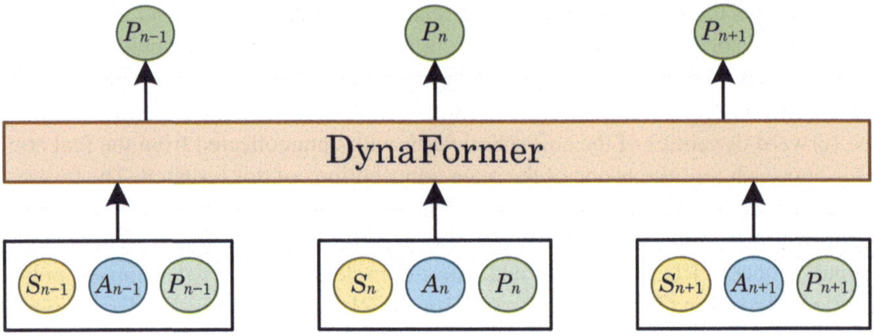


**Fig. 1** The structure of the DynaFormer model with hashed tokens are masked tokens

## *2.1 The DynaFormer*

The underlying structure of the DynaFormer adheres to the recent LLaMA model [17]. The LLaMA model in turn follows the decoder of the original transformer with certain modifications. These modifications include the usage of Root Mean Square Layer Normalization (RMSNorm) instead of the Layer Normalization and the utilization of a different activation function for the feed forward layer. The architectural layout of the DynaFormer model is shown in the Fig. 2. The Layer Normalization [18] is designed to normalize the hidden states of the neural network, thereby aiding in the prevention of vanishing or exploding gradients during the training process. The norm layer re-centers the hidden states by subtracting the mean of the states and rescaling by dividing with the standard deviation. On the other hand, RMSNorm solely rescales the hidden states by dividing each element in the hidden states by the root mean square value of the entire hidden states [19]. The RMS norm is found to have almost the same effect as the norm layer. In addition, the RMS norm is computationally simpler, which makes the training faster. The activation function used in the feedforward neural network is the SwiGLU function [20]. The SwiGLU function excels over other possible activation functions and, therefore, is used in the proposed DynaFormer model.

The DynaFormer model incorporates an initial embedding layer that individually learns the embeddings for each element in the input $[S_n, A_n, P_n]$. This layer is defined as input embedding. The dimension $d$ of the embedding is a hyper-parameter. Transformer models process the input sequence in parallel. In order to determine the position of each element within the input sequence, a positional encoding layer is introduced to the input embedding. The model comprises of two sub-layers, each preceded by a pre-normalization layer. The first sub-layer is a masked self-attention layer, and the second sub-layer is a feedforward network. These two sub-layers constitute a block. The number of blocks in the model is another hyper-parameter. The masked self-attention sub-layer enables the model to take into account the relationships between different elements in the input sequence,

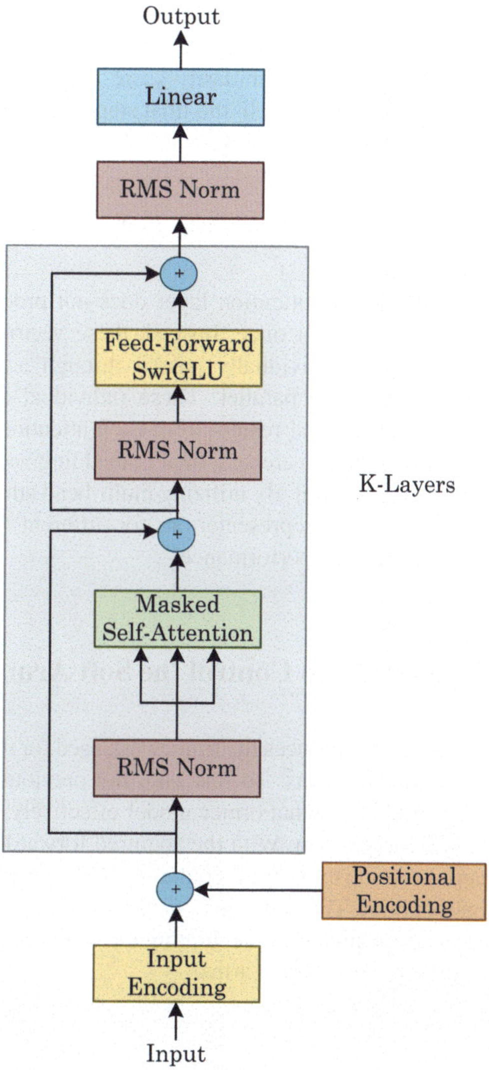


**Fig. 2** The internal architecture of the DynaFormer model

even if they are far apart. This sub-layer is also what gives the transformer the ability to effectively process long sequences. The mask is used with self-attention to control which positions in the input sequence are attended to. It limits the model to consider only the elements from the start of the sequence until time step $n$.

After passing through the input embedding and positional encoding, each member $[S_n, A_n, P_n]$ in element $\tau_n$ of the input sequence is stacked together to create new element. This new element in the sequence is the input to three neural networks with one layer each. These networks generate three vectors, specifically the Query (**Q**), the Key (**K**), and the Value (**V**) vectors. The dimension $d$ of these three vectors is identical to the size of the output of the embedding and positional encoding layers. The Query vectors are multiplied by the Key vectors to obtain a similarity score. The similarity score is then subjected to a scaling of $\sqrt{d}$ and a softmax normalization. In the final step, the normalized and scaled similarity score is multiplied by the Value vectors. Thus, the attention is computed as

$$\text{Attention}(\mathbf{Q}, \mathbf{K}, \mathbf{V}) = \text{softmax}\left(\frac{\mathbf{Q} \cdot \mathbf{K}^T}{\sqrt{d}}\right)\mathbf{V} \tag{1}$$

In general, the attention layer does not process the complete Query, Key, and Value vectors all at once. Instead, these vectors are divided into $M$ groups, and each group is individually passed through a separate instance of the attention layer operating in parallel. These individual attention layers are referred to as attention heads. The results from each attention head are then combined together by concatenation to create a final embedding vector. This approach is referred to as multi-head attention. By utilizing multi-head attention, the transformer is capable of acquiring distinct representations for different segments of the sequence, which is shown to enhance performance.

## 3 Learning to Control the Soft Arm

This section introduces the framework used for the autonomous learning and control of soft robotic arms. To this end, the previously introduced DynaFormer model is utilized. The DynaFormer model effectively captures the complex dynamics of the soft robotic arm. With the acquired forward dynamics model, the twin-delayed deep deterministic policy gradient (TD3) algorithm [16] is employed to learn the control policy. This innovative methodology bridges the gap between simulation and real-world application, facilitating the development of robust and adaptable control policies for soft robotic arms.

### *3.1 Twin-Delayed Deep Deterministic Policy Gradient Algorithm*

Reinforcement learning (RL) is a part of machine learning that enables autonomous learning of agents through interaction with an environment. In RL, an agent aims to learn a policy that maximizes a cumulative reward signal while exploring and interacting with the environment that is modeled as a Markov decision process (MDP). An MDP can be defined as a tuple $< \mathcal{S}, \mathcal{A}, \mathcal{P}, R, \gamma >$ where $\mathcal{S}$ represents the set of all possible states that the agent can attain. These states consist of the goal to reach in addition to the state of the arm. The set of actions representing the choices the agent can make is represented by $\mathcal{A}$, while $\mathcal{P}(s'|s, A)$ is the transition probability function defining the probability of transitioning from state $s \in \mathcal{S}$ to state $s' \in \mathcal{S}$ after taking action $A \in \mathcal{A}$. Note that the RL agent's state is denoted using lowercase $s$ to distinguish it from the arm's state $S$ in the DynaFormer model, while the action $A$ is common to both methods. $R(s, A, s')$ is the reward function specifying the immediate reward received after transitioning from state $s$ to $s'$ by taking action $A$. The reward $R$ is determined by the distance between the tip of the arm and the goal, with a negative value indicating close proximity to the goal. Finally, $\gamma$ is the discount factor determining the weight given to future rewards.

The goal of RL is to find a policy $\pi(A|s)$ that maximizes the expected cumulative reward, often quantified by the expected return $J(\pi)$:

$$J(\pi) = \mathbb{E}\left\{\sum_{n=0}^{\infty} \gamma^n R(s_n, A_n, s_{n+1})\right\} \tag{2}$$

where $\mathbb{E}\{\cdot\}$ is the expectation operator.

When it comes to practical implementation such as robotics control, RL algorithms may encounter difficulties when used with action spaces that are both high dimensional and continuous. A notable algorithm that tackles this particular challenge is the TD3 algorithm. The TD3 algorithm is an off-policy, model-free RL algorithm designed for continuous action spaces. TD3 follows the actor-critic framework which contains two main components, namely, the actor network and the critic network. The actor network represents the policy that maps the given state into an action. The critic network takes the chosen action with the given state as input and estimate the Q-value to update the policy. TD3 builds on the original deep deterministic policy gradient (DDPG) algorithm [10] and addresses some of its limitations such as overestimation of Q-values and instability during training. TD3 employs two Q-value networks, target networks, and a target policy smoothing technique to enhance training stability. The key equations of the TD3 algorithm are as follows:

1. Target Q-Value Network Update:

$$\begin{aligned} \hat{Q}_1(s, A) &= R(s, A, s') + \gamma \cdot \min(Q_{1\text{-target}}(s', \pi_{\text{target}}(s')) \\ \hat{Q}_2(s, A) &= R(s, A, s') + \gamma \cdot \min(Q_{2\text{-target}}(s', \pi_{\text{target}}(s')) \end{aligned} \tag{3}$$

2. Critic Loss:

$$\mathcal{L}_{\text{critic}} = \frac{1}{2}\mathbb{E}\left\{(Q_1(s, A) - \hat{Q}_1(s, A))^2 + (Q_2(s, A) - \hat{Q}_2(s, A))^2\right\} \tag{4}$$

3. Actor Loss:

$$\mathcal{L}_{\text{actor}} = -\mathbb{E}\{Q_1(s, \pi(s))\} \tag{5}$$

4. Target Network Update:

$$\begin{aligned} Q_{1\text{-target}}(s, A) &= (1-\tau)Q_{1\text{-target}}(s, A) + \tau Q_1(s, A) \\ Q_{2\text{-target}}(s, A) &= (1-\tau)Q_{2\text{-target}}(s, A) + \tau Q_2(s, A) \\ \pi_{\text{target}}(s) &= (1-\tau)\pi_{\text{target}}(s) + \tau\pi(s) \end{aligned} \tag{6}$$

where $Q_1(s, A)$ and $Q_2(s, A)$ are the Q-value networks, $\pi(s)$ is the policy network, and $\tau$ is a soft target network update coefficient.

## 3.2 The Learning Framework

The initial phase in constructing the framework to learn the control policy for a robotic arm entails building the DynaFormer model. In order to train the DynaFormer model, data is gathered in the format of $[S, A, P]$, where $S$ is the state of the arm, $A$ denotes the action, and $P$ represents the position of the arm's tip. The data is collected in constant time steps. The action $A$ is taken at the beginning of each time step, and the resultant tip position $P$ is recorded at the end of the time step. The trained DynaFormer model captures the dynamics of the arm and is utilized as a simulation environment to facilitate the training of the control policy via the employment of the RL TD3 algorithm. Figure 3 illustrates the system employed for training the RL agent responsible for controlling the flexible arm.

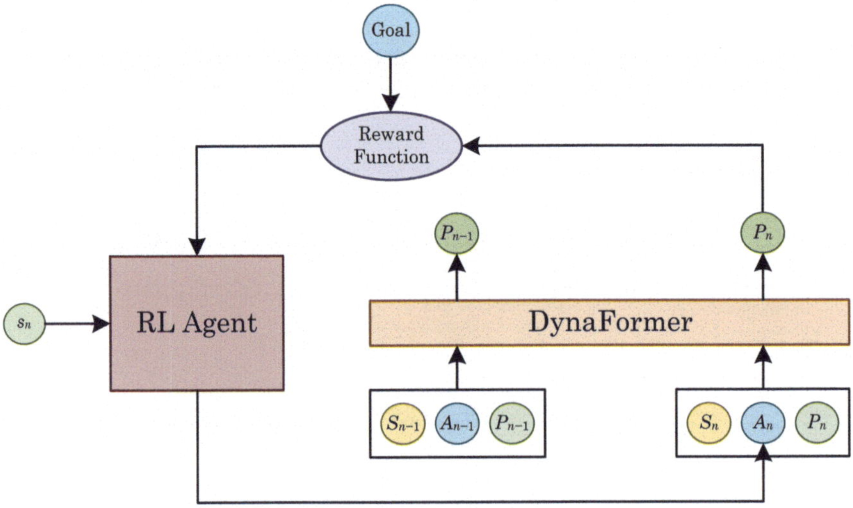


**Fig. 3** The learning framework of the control policy for the robotic arm

# 4 Experiments

This section presents the experimental setup, testing, and results of the framework for learning to control a soft robotic arm using RL. The experiment combines learning the DynaFormer model for obtaining the forward dynamics of the robotic arm and learning the control policy for a soft robotic arm using TD3 algorithm.

## *4.1 Soft Robotic Arm*

The robotic arm used to test the learning framework is a soft arm model (see Fig. 4) that is implemented in the Simulation Open Framework Architecture (SOFA) software [21]. SOFA is an open-source framework designed for physics simulation. The software was developed in particular for modeling and simulation of soft and deformable materials. It allows creating and customizing simulation models for various applications such as tissue manipulation and soft robots. The SOFA simulation is based on the finite element method (FEM) used to model the nonlinear deformation of soft structures, actuators, and surroundings. The model of the arm was first introduced by Coevoet et al. [22]. This soft robotic arm (named Trunk) is actuated by eight cables. Four cables actuate the first section of the arm and four go through the entire arm length. The length of the arm is 195 mm. The arm is controlled through cable displacements with a maximum displacement of 70 mm for the cables that go through the entire arm and 40 mm for the cables that control only the first section of the arm.

## *4.2 Learning The DynaFormer Model*

The training process of DynaFormer consists of two main steps, namely, data collection and model training. The default time step of the SOFA program is 10 ms. At each time step, data is collected as follows. First the displacement of the cables $A_{n-1}$ and the current tip position $P_{n-1}$ are recorded to form the state $S_n$. Then, each cable is pulled or released by a maximum of 1 mm (representing $A_n$) and the resultant tip position $P_n$ is recorded.

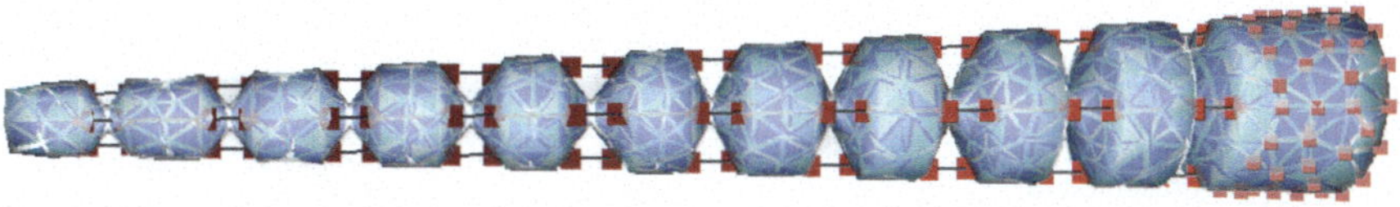

**Fig. 4** The soft robotic arm used to test the learning framework

The DynaFormer model consists of 12 blocks and 8 attention heads. The length of the hidden dimension used for the model is 128. As a baseline, a LSTM model is trained to benchmark the DynaFormer model. The LSTM model utilizes past three time steps where the input is the state of the arm $S_n$ and the action $A_n$ taken at that time step. These are same states $S_n$ and actions $A_n$ used for the DynaFormer model. The output of both the LSTM model and DynaFormer is the tip position $P_n$. Both models are trained on NVIDIA GeForce 2060 GPU. The LSTM model is trained for 100 epochs, while the DynaFormer model is trained for 300 epochs. This difference in number of epochs is due to the larger and more comprehensive model utilized by the DynaFormer. The mean square error (MSE) serves as the loss function for training both models. The models are trained using the AdamW optimizer [23] with a learning rate of $10^{-4}$ and $10^{-2}$ for the DynaFormer and LSTM models, respectively. As the cables are pulled or released by 1 mm at each time step, the tip of the arm moves with a maximum of 7.78 mm along the $x$- and $y$-axes, and by 9.7 mm along the $z$-axis, with mean and standard deviation of $2.8 \pm 2.1$ mm at the $x$- and $y$-axes, and $3.4 \pm 1.8$ mm about the $z$-axis. The following Table 1 shows the comparison between the mean absolute error (MAE) and standard deviation (STD) of the two models on predicting the tip position.

From these results, it is clear that DynaFormer and LSTM demonstrate distinct performance characteristics. DynaFormer consistently outperforms the LSTM model and achieves lower MAE values with narrower ranges of standard deviation, indicating more consistent and precise predictions. The MAE of the DynaFormer model is around 6%, 2.5%, and 2.3% of the mean movements of the tip along the $x$-, $y$-, and $z$-axis, respectively. In comparison, the LSTM MAE is around 8.2%, 9.6%, and 5.8% of the mean movements of the tip. These findings suggest that DynaFormer is a more effective model for learning forward dynamics in this specific context, as it consistently generates more accurate predictions with less variability.

### *4.3 Learning The Control Policy*

The control policy is learned using the TD3 algorithm. Being a deterministic algorithm, TD3 generates the same action for a given specific state. Therefore, initializing with a specific starting state will generate the same set of actions, preventing the training the agent. To circumvent this problem, a normally distributed noise signal with a zero mean and a standard deviation of 0.1 is added to the output action (whose value is between -1 and 1) to ensure that new actions are explored.

**Table 1** The MAE and STD of predicting the tip position. All results are presented in mm

| Model | $x$-axis | $y$-axis | $z$-axis |
|---|---|---|---|
| LSTM | $0.23 \pm 0.58$ | $0.27 \pm 0.57$ | $0.20 \pm 0.52$ |
| DynaFormer | $0.17 \pm 0.03$ | $0.07 \pm 0.02$ | $0.08 \pm 0.03$ |

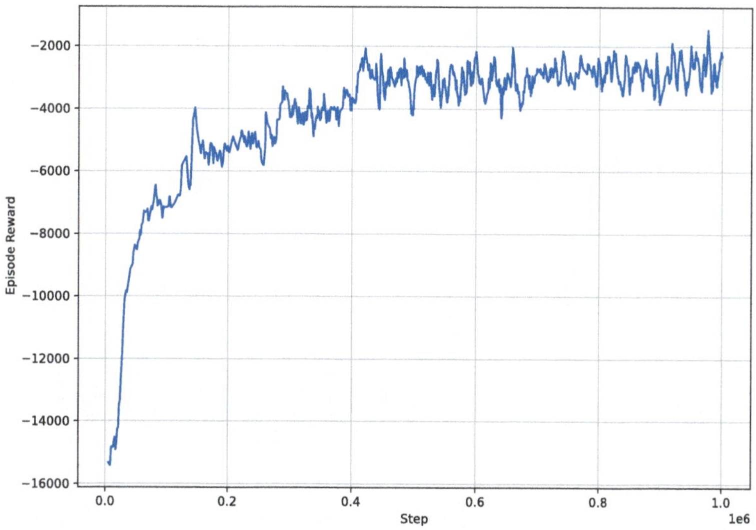


**Fig. 5** The evolution of the reward during the agent training

This procedure (referred to as exploration-exploitation) is utilized in this work for the training of the RL agent.

The replay buffer used during training the TD3 RL agent is Hindsight Experience Replay (HER) [24]. HER is a technique that enables the agent to learn from unsuccessful outcomes by replaying experiences with different goal states, improving its learning efficiency and ability to tackle diverse scenarios. The evolution of both the reward and the episode length throughout the training process is plotted in Figs. 5 and 6. This graphical representation provides a clear insight into the agent's learning dynamics and performance improvements over time. Figure 5 shows the reward evolution during the agent training, demonstrating the agent's ability to maximize its rewards as the training proceeds. Figure 6 represents the episode length evolution during the agent training. This figure shows that as the learning proceeds, the episode length decreases. Such a decrease in episode length is an indication that the agent can reach the goals faster with time. The episode length during the training is set to 150 steps. The average episode length of a trained agent is in the range of 50 steps. This indicates the agent's ability to reach the goal in a short amount of time.

## *4.4 Results*

The workspace utilized for training the dynamic model and RL agent (see Fig. 7) was designed to closely replicate real-world scenarios. The goal of the RL agent is to reach the sampled goals from the arm's workspace. To evaluate the accuracy of the RL agent's performance, an assessment criterion is employed. When the tip of

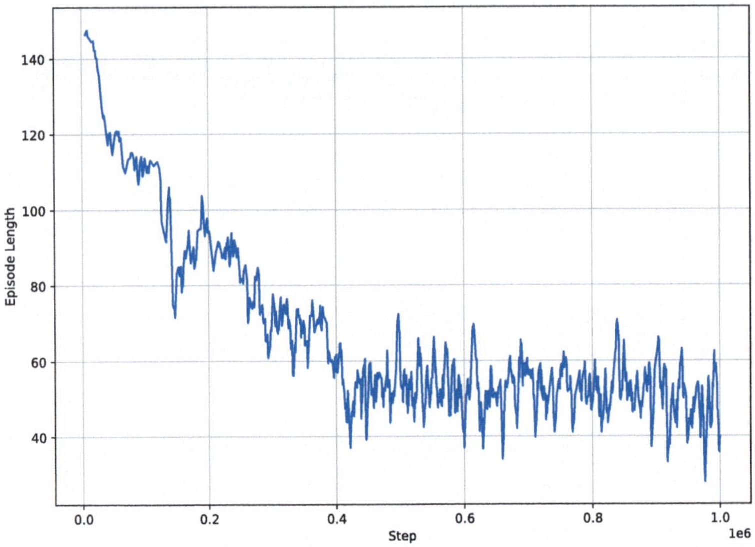


**Fig. 6** The evolution of the episode length in the form of time steps during the agent training

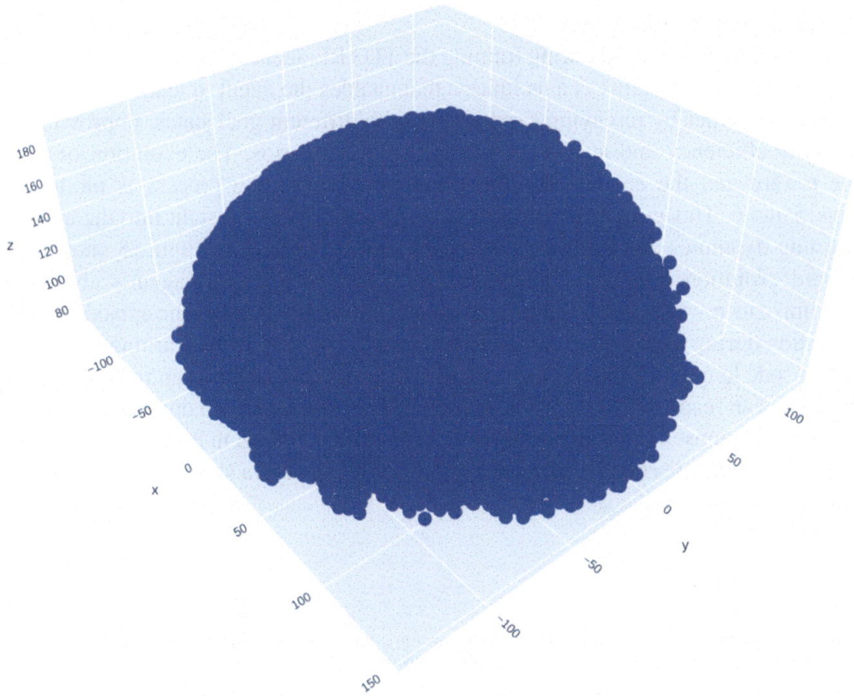


**Fig. 7** The workspace of the arm used in the training of both the dynamic model and the RL agent

the arm reaches within 10 mm of the goal, it is a considered a success for the agent. The goal is a ball with a radius of 5 mm. During the RL agent training, the length of each episode spanned 150 time steps. At the end of the training, the average length of the episode is around 50 time steps for the trained agent. When tested, the trained agent is able to reach the goals with 100% accuracy using episodes of length 150 steps.

## 5 Conclusion

This paper showcases a comprehensive framework that utilizes reinforcement learning (RL) to control a soft robotic arm. The framework includes a forward dynamics model called DynaFormer, which is based on transformer architecture. The DynaFormer is designed specifically to capture the dynamic behavior of soft and deformable materials. Once the forward model is obtained, it is used to train the RL model for positioning the soft robot arm tip. Simulation-based results highlight the effectiveness and potential of this approach in the field of soft robotics. The learning framework employs the SOFA platform to test its capabilities. This research contributes to the continuous advancement of soft robotics and its applications in various domains, ranging from medical to industrial settings. The fusion of RL and the modeling of soft and deformable materials holds immense potential for the development of highly flexible and adaptable robotic systems. One limitation of the proposed approach resides in its computational expenses. Transformer-based architectures such as DynaFormer necessitate substantial computational resources for both training and inference. Consequently, this could impede the effective training of the RL agent. Another limitation is related to the data requirements of transformer models. These models demand a copious amount of data for effective training, making them difficult to implement on physical soft arms. This problem can be alleviated by pretraining a model using simulated data, prior to fine-tuning the model on actual data obtained from the arm. Future research is going to be directed toward studying real-world implementations of learning to control soft robotic arm in practical scenarios. This research will further validate the framework's potential for control applications.

## References


1. Holsten, F., Engell-Norregard, M.P., Darkner, S., Erleben, K.: Data driven inverse kinematics of soft robots using local models. In: IEEE Int. Conf. Robot. Autom., pp. 6251–6257, Montreal, Canada (2019)
2. Bern, J.M., Schnider, Y., Banzet, P., Kumar, N., Coros, S.: Soft robot control with a learned differentiable model. In: 3rd IEEE Int. Conf. Soft Robotics, pp. 417–423, New Haven, CT (2020)

3. Gillespie, M.T., Best, C.M., Townsend, E.C., Wingate, D., Killpack, M.D.: Learning nonlinear dynamic models of soft robots for model predictive control with neural networks. In: IEEE Int. Conf. Soft Robotics, pp. 39–45, Livorno, Italy (2018)
4. Kovandzic, M., Nikolić, V., Simonovic, M., Ćiric, I., Al-Noori, A.: Soft robot positioning using artificial neural network. Facta Universitatis, Series: Automatic Control and Robotics, vol. 18(1), pp. 19–30 (2019)
5. Ansari, Y., Manti, M., Falotico, E., Cianchetti, M., Laschi, C.: Multiobjective optimization for stiffness and position control in a soft robot arm module. IEEE Robot. Autom. Lett. **3**, 108–115 (2018)
6. Zhang, H., Cao, R., Zilberstein, S., Wu, F., Chen, X.: Toward effective soft robot control via reinforcement learning. In: 10th International Conference on Intelligent Robotics and Applications (ICIRA), Part I 10, pp. 173–184, Wuhan, China (2017)
7. Satheeshbabu, S., Uppalapati, N.K., Chowdhary, G.V., Krishnan, G.: Open loop position control of soft continuum arm using deep reinforcement learning. In: International Conference on Robotics and Automation (ICRA), pp. 5133–5139 (2019)
8. Mnih, V., Kavukcuoglu, K., Silver, D., Graves, A., Antonoglou, I., Wierstra, D., Riedmiller, M.A.: Playing Atari with deep reinforcement learning. ArXiv: 1312.5602 (2013)
9. Satheeshbabu, S., Uppalapati, N.K., Fu, T., Krishnan, G.: Continuous control of a soft continuum arm using deep reinforcement learning. In: 3rd IEEE International Conference on Soft Robotics (RoboSoft), pp. 497–503 (2020)
10. Lillicrap, T. P., Hunt, J. J., Pritzel, A., Heess, N. M., Erez, T., Tassa, Y., Silver, D., Wierstra, D.: Continuous control with deep reinforcement learning. ArXiv: 1509.02971 (2015)
11. Centurelli, A., Arleo, L., Rizzo, A., Tolu, S., Laschi, C., Falotico, E.: Closed-loop dynamic control of a soft manipulator using deep reinforcement learning. IEEE Robot. Autom. Lett. **7**(2), 4741–4748 (2022)
12. Chen, L., Lu, K., Rajeswaran, A., Lee, K., Grover, A., Laskin, M., Abbeel, P., Srinivas, A., Mordatch, I.: Decision transformer: Reinforcement learning via sequence modeling. ArXiv: 2010.11929 (2021)
13. Radford, A., Narasimhan, K.: Improving language understanding by generative pre-training, unpublished. https://s3-us-west-2.amazonaws.com/openai-assets/research-covers/language-unsupervised/language_understanding_paper.pdf. Last access date: 30 Oct. 2022
14. Vaswani, A., Shazeer, N., Parmar, N., Uszkoreit, J., Jones, L., Gomez, A. N., Kaiser, L., Polosukhin, I.: Attention is all you need. In: Advances in Neural Information Processing Systems (NeurIPS), vol. 30 (2017)
15. Esslinger, K., Platt, R.W., Amato, C.: Deep transformer Q-networks for partially observable reinforcement learning. ArXiv: 2206.01078 (2022)
16. Fujimoto, S., Hoof, H.V., Meger, D.: Addressing function approximation error in actor-critic methods. In: Proc. Int. Conf. on Machine Learning (PMLR), pp. 1587–1596 (2018)
17. Touvron, H., Lavril, T., Izacard, G., Martinet, X., Lachaux, M., Lacroix, T., Rozière, B., Goyal, N., Hambro, E., Azhar, F., Rodriguez, A., Joulin, A., Grave, E., Lample, G.: LLaMA: Open and efficient foundation language models. ArXiv: 2302.13971 (2023)
18. Ba, J., Kiros, J.R., Hinton, G.E.: Layer normalization. ArXiv: 1607.06450 (2016)
19. Zhang, B., Sennrich, R.: Root mean square layer normalization. ArXiv: 1910.07467 (2019)
20. Shazeer, N.M.: GLU variants improve transformer. ArXiv: 2002.05202 (2020)
21. Faure, F., Duriez, C., Delingette, H., Allard, J., Gilles, B., Marchesseau, S., Talbot, H., Courtecuisse, H., Bousquet, G., Peterlik, I.: Sofa: A multi-model framework for interactive physical simulation. Studies in Mechanobiology Tissue Engineering and Biomaterials, pp. 283–321 (2012)
22. Coevoet, E., Escande, A., Duriez, C.: Optimization-based inverse model of soft robots with contact handling. IEEE Robot. Autom. Lett. **2**(3), 1413–1419 (2017)
23. Loshchilov, I., Hutter, F.: Decoupled weight decay regularization. ArXiv: 1711.05101 (2017)
24. Andrychowicz, M., Crow, D., Ray, A., Schneider, J., Fong, R., Welinder, P., McGrew, B., Tobin, J., Abbeel, P., Zaremba, W.: Hindsight experience replay. Adv. Neural Inf. Process. Syst., 30 (2017)

# KineFormer: Solving the Inverse Modeling Problem of Soft Robots Using Transformers

**Abdelrahman Alkhodary and Berke Gur**

## 1 Introduction

As an alternative to rigid electromechanical manipulators, hyper-elastic soft manipulators have gained significant research interest in the past decade. Compared to their electromechanical counterparts, soft manipulators generally provide low performance at a low cost. However, soft manipulators make up for this low performance through their compliance, eliminating the need for high-precision motion and force control strategies, while making them very suitable for operating in fragile environments.

Unfortunately, due to the continuum nature of these soft manipulators, developing accurate forward analytical models that map the actuator space to the task space is challenging, making inverse analytical modeling intractable. Although established methods developed for conventional rigid manipulators can be modified for modeling soft and deformable robots, these methods are approximate and have limited applicability. Therefore, developing effective and efficient inverse models for soft manipulators remains an elusive problem.

With this paper, a transformer architecture-based approach for machine learning the inverse kinematics of soft robotic arms is presented. To the best knowledge of the authors, this paper represents the first attempt to use the transformer

This research was supported by The Scientific and Technological Research Council of Türkiye (TÜBİTAK) through the project entitled "Robotic Squid for Underwater Manipulation and Intervention" (grant no. 216M201). The funder had no role in study design, data collection and analysis, decision to publish, or preparation of the manuscript.

A. Alkhodary (✉) · B. Gur
Bahcesehir University, İstanbul, Türkiye
e-mail: abdelrahman.alkhoda@bahcesehir.edu.tr; berke.gur@bau.edu.tr



Ö. M. Gül et al. (eds.), *7th EAI International Conference on Robotic Sensor Networks*, EAI/Springer Innovations in Communication and Computing,
https://doi.org/10.1007/978-3-031-64495-5_3

architecture-based machine learning methods for shape and motion control of soft robots. The remainder of the paper is organized as follows. A concise review of existing literature on kinematic modeling of soft robotic arm models is presented in Sect. 2. This is followed by Sect. 3 where the proposed inverse model based on the transformer architecture, namely, the Kinematics Transformer (or KineFormer, for short), is described. Numerical evaluation of the KineFormer in the form of simulations is described in Sect. 4. Results pertaining to the numerical evaluation of the KineFormer are presented and discussed in Sect. 5. The paper is concluded by outlining future work in Sect. 6.

## 2 Related Work

The finite element method (FEM) is a numerical technique used to simulate the behavior of deformable structures. FEM discretizes the structure into a mesh. Each mesh element is then analyzed in terms of its geometry, material properties, and boundary conditions to obtain a solution to the governing equations of the problem. Because soft robots are made from highly deformable materials, FEM is largely used to model, simulate, and study such robots. In [1], the authors proposed a kinematic model and tracking control method for soft robots based on FEM. A closed-loop controller is designed with a position-based visual servoing strategy. In addition, a novel switched control method is proposed to handle image feature loss. Frederick et al. presented a high-rate control solution for soft robots that uses finite element analysis to compute an inverse model of the robot. This model outputs set-point values for actuators to achieve a desired trajectory [2]. A quadratic programming algorithm based on FEM is used to solve the inverse problem, and an asynchronous simulation framework is proposed to improve update rate performance.

One very important limitation of FEM is that solving complex and dynamic problems can be computationally expensive, hindering its utilization in real-time applications such as tele-operation. In addition, FEM requires precise knowledge of the material properties, boundary conditions, and actuator characteristics of the soft robot. These parameters may not always be known or easily measured. In contrast, machine learning techniques can be exploited to develop numerical models directly from experimental data collected from the actual robot, thus eliminating the requirement to determine material and topological parameters and avoid modeling errors. Once trained, machine learning techniques are computationally more efficient, potentially enabling soft or even hard real-time operation.

Holsten et al. learned the inverse kinematic model of a soft robot arm using polynomial regression. In this work, the input is the shape of the arm and the output is the cable displacements that control the soft arm [3]. The shape of the arm is represented as points distributed over the soft arm. Gillespie et al. trained a neural network to learn the nonlinear dynamics of a soft pneumatic arm [4]. This learned nonlinear model Jacobians are then used to create a linear state-space model of the soft arm. This linearized model is used together with a model predictive controller for motion control. In [5], the forward kinematic model of a soft robotic arm was learned either using data obtained from the actual arm or from data computed using a FEM of the arm. Once this relatively simple forward kinematic model is obtained, the inverse model is computed from this forward neural network model by minimizing a cost function (defined in terms of the error of the tip position) using the gradient descent optimization algorithm through the neural network Jacobians.

The authors of [6] investigated the utilization of neural networks for modeling the inverse kinematics of a soft arm. The soft arm used in this research has three identical hollow cylinders, each with a rubber hose passing through them. The arm is pneumatically actuated. The inverse kinematic model learns the mapping between the amount of pressure applied to each rubber hose and the position of the tip of the arm. In [7], Thuruthel et al. attempt to learn the inverse kinematic model of a continuum robot. The data used to train the inverse neural network model was generated from a kinematic model that uses a constant curvature approximation for modeling the continuum arm. The arm consists of three segments, with each segment controlled by three pneumatic actuators. The input to the inverse model is the current actuation of the segments and the desired tip position of the arm. The output is the actuation of the segment to reach the desired tip position.

The feed-forward neural networks (FFNN) are designed to learn input-output mappings. Therefore, these networks are very suitable for learning tasks such as regression. For example, in learning the inverse kinematics of robotic arms, FFNN's are used to map the desired tip position to the corresponding actuation inputs. However, FFNN's cannot keep track of sequential data and thus are not very suitable for learning sequential problems such the time-dependent inverse dynamic modeling of the arm motion.

The inverse kinematics problem can be cast into a sequential decision-making problem by segmenting the path between the current state (e.g., tip position) and the final desired state into intermediate waypoints. As stated above, the inverse dynamics problem involves modeling time evolution of the trajectory of the arm and therefore, by definition, is sequential. For such sequential problems, alternative architectures and algorithms that stem from other areas of machine learning such as natural language processing are more suitable. This paper proposes an architecture called Kinematics Transformer (KineFormer) to solve the inverse kinematics problem based on the transformer architecture. The KineFormer model can be used for solving the inverse kinematics problem of reaching a single desired position or multiple waypoints since it considers past positions along the trajectory.

The transformer architecture was first introduced by Vaswani et al. [8] for language translation. The core of the transformer architecture is the attention mechanism. The transformer architecture has an encoder and decoder parts, while the encoder has self-attention mechanism, and the decoder has mask self-attention mechanism. Most recent state-of-art models in natural language processing are based on the transformer architecture. One such model, namely, the BERT architecture [9], is based on the encoder part of the transformer. This method is used for question answering, where a sentence is fed to the architecture with a mask token for the missing word. Generative pretrained transformers (GPT) are architectures based on the decoder part of the transformer and, as the name implies, are used to generate an output based on the input sequence [10]. Transformer architecture is also proposed in [11] for image classification (dubbed the vision transformer). Because transformer architecture can deal with sequences of data and not images, the authors of vision transformer proposed to crop the input image into a set of patches. Each patch is flattened and linearly projected into a fixed dimension. Chen et al. [12] proposed a new framework, namely, the decision transformer that casts reinforcement learning as a conditional sequence modeling problem and utilizes the transformer architecture for the solution.

## 3 Learning the Kinematics of Soft Robots

The primary contribution of this paper, namely, the KineFormer, is presented in this section. The KineFormer is a deep learning model based on the GPT architecture, with the primary objective of modeling the inverse kinematics of soft robotic arms. Similar to other transformer-based architectures, the KineFormer also has the so-called attention mechanism in its core.

### 3.1 Inverse Kinematics as a Sequence

The primary motivation for developing the KineFormer was to effectively learn the inverse kinematics of soft robotic arms in reaching a single point in 3D space or reaching multiple points in a sequential order. Such a sequence is represented with $N$-element list $\boldsymbol{\Sigma} = [\boldsymbol{\tau}(0), \boldsymbol{\tau}(1), \cdots, \boldsymbol{\tau}(N-1)]$ where each element in the list is defined as $\boldsymbol{\tau}(n) = [\mathbf{s}(n), \mathbf{r}_d(n), \mathbf{A}_d(n)]$ for $0 \leq n \leq N-1$. The state of the arm is expressed as $\mathbf{s}(n) = [\mathbf{A}(n), \mathbf{r}(n)]$ with $\mathbf{A}(n)$ and $\mathbf{r}(n)$ representing the actuation values and the tip position, respectively. As shown in Fig. 1, other components of the element include the goal or desired tip position $\mathbf{r}_d(n)$ and the desired (but unknown) actuation values $\mathbf{A}_d(n)$. The desired actuation values $\mathbf{A}_d(n)$ are masked with zero and are predicted by the model. These predicted actuation values are applied to the arm and become the actual actuation values in the next step, i.e., $\mathbf{A}(n+1) = \mathbf{A}_d(n)$. Note that the current desired tip position $\mathbf{r}_d(n)$ can be different from the next actual tip position $\mathbf{r}(n+1)$ due to errors in predicting the desired actuation values.

### 3.2 The KineFormer

The base architecture of the KineFormer is the GPT model. GPT is a $k$-layer decoder-only transformer. Each decoder layer is comprised of a masked multi-head self-attention mechanism as the first sub-layer and a position-wise fully connected

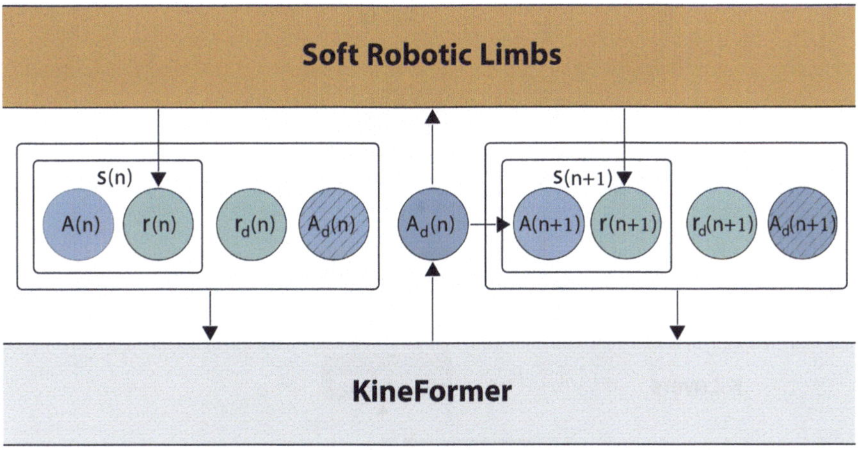


**Fig. 1** The element structure used in the KineFormer with masked elements shown with a pattern fill

second sub-layer, as shown in Fig. 2 [10]. GPT models are auto-regressive models in the sense that they generate a new element based on past input elements of the sequence. The auto-regressive characteristic of the model comes from the masked self-attention layer. Each element in $[\mathbf{s}(n), \mathbf{r}_d(n), \mathbf{A}_d(n)]$ first passes through a single layer neural network to learn the embeddings for each element. This layer is referred to as the input embedding. As is discussed in the next section, the transformer architecture processes the entire input sequence in parallel using the self-attention mechanism. Therefore, without any notion of position, the model would treat each of the elements of the input sequence independently and lose any sequential information. To mitigate this problem, another single layer neural network (termed the positional encoding) is used to learn the positional encoding for each element in the sequence [12].

### *3.3 Self-Attention*

The masked self-attention mechanism is one of two sub-layers of the GPT transformer architecture. The mask is used to predict the element at step $n$ limits the GPT to consider only the elements from the start of the sequence until step $n$, including $\mathbf{s}(n)$ and $\mathbf{r}_d(n)$ at the step $n$ while masking $\mathbf{A}_d(n)$. The self-attention enables the model to take into account the relationships between different elements in the input sequence, even if they are far apart. The output of this masked self-attention sub-layer is a contextually richer representation of the input sequence. This sub-layer is also what gives the transformer the ability to effectively process long sequences.

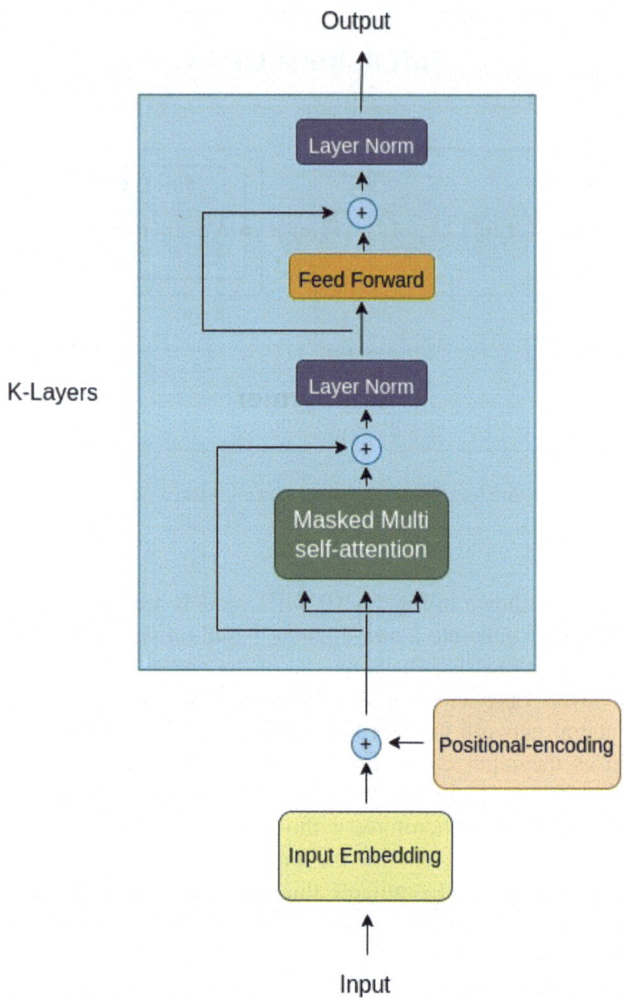


**Fig. 2** The architecture used in the KineFormer

After passing through the embedding and positional encoding layers, the state $\mathbf{s}(n)$, the desired tip position $\mathbf{r}_d(n)$, and the actuation values $\mathbf{A}_d(n)$ are stacked together and treated as one element. Each element in the sequence is fed to three one-layer neural networks that generates three vectors, namely, the Query (**Q**), the Key (**K**), and the Value (**V**) vectors. The dimension $d$ of these three vectors is the same as the dimension of the output of the embedding and positional encoding layers. This embedding (or hidden) dimension $d$ is a hyper-parameter that is determined by the user. The Query vectors are multiplied by the Key vectors to get a similarity score. A scaling of $\sqrt{d}$ and a softmax normalization is applied to the similarity score. In the final step, the normalized and scaled similarity score is multiplied by the Value vectors. Thus, the attention is computed as

$$\text{Attention}(\mathbf{Q}, \mathbf{K}, \mathbf{V}) = \text{softmax}\left(\frac{\mathbf{Q} \cdot \mathbf{K}^T}{\sqrt{d}}\right)\mathbf{V} \tag{1}$$

In general, the complete Query, Key, and Value vectors are not passed through the attention layer all at once. Instead, these vectors are split into $M$ groups, and each group is fed into a separate instance of the attention layer, running in parallel. These attention layers are called attention heads. Attention results from each attention head are concatenated together to produce a final embedding vector. This process is termed multi-head attention. Using multi-head attention allows the transformer to learn different representations of the parts of the sequence which is known to increase the performance.

The KineFormer can also operate without the state element $\mathbf{s}(n)$. However, during development, it was determined that the learning performance was improved by incorporating the state into the inputs. This improvement is attributed to the fact that the state element provides the transformer with information regarding the current configuration of the arm, which is exploited in estimating the necessary inputs for the next tip position in the sequence.

## 4 Simulations

This section provides results pertaining to the numerical evaluation of the proposed method. To prove its versatility, the KineFormer is tested using two very different soft robot models. The first soft robotic arm has an analytic forward kinematic model developed based on the Cosserat theory. The second arm model is based on the FEM and does not have an assumed closed forward kinematics solution. Both arms are tendon actuated.

### *4.1 Cosserat Arm*

Inspired by the arms and tentacles of squid, a slender solid frustum-shaped geometry is assumed for the arms. The Cosserat theory is adapted for developing analytical models of the arms. In this section, details regarding the analytic arm model used in this study are provided.

#### 4.1.1 Analytical Arm Model

Following the approach presented by Ren et al., an analytic kinematic model for the arms is developed based on the Cosserat theory [13]. Accordingly, the shape of a arm of nominal length $L$ is described using the material arc length coordinate $0 \leq s \leq L$ that passes through the centroids of the circular cross sections along the longitudinal axis of the arm. In addition to an inertial coordinate system fixed to the base of the arm (with unit vectors $\mathbf{D}_1$, $\mathbf{D}_2$, and $\mathbf{D}_3$), local coordinate systems attached to the cross sections are used to describe the state of arm in the task space in terms of position $\mathbf{r}(s)$ and orientation $\mathbf{R}(s)$. These local coordinate systems are defined by the two principle axes (with corresponding unit vectors $\mathbf{d}_1$ and $\mathbf{d}_2$) as well as an axis perpendicular to the cross section (with a corresponding unit vector $\mathbf{d}_3$).

The state of the arm in the configuration space is described in terms of the longitudinal strain $\epsilon_3(s)$ and three curvatures $\boldsymbol{\kappa}(s) = [\kappa_1(s)\ \kappa_2(s)\ \kappa_3(s)]^T$, defined about the corresponding local coordinate axes. In this model, the Bernoulli-Euler beam assumption is exploited, where the cross sections are assumed to remain in-plane, making the shear strains $\epsilon_1$ and $\epsilon_2$ equal to zero. The relationship between the longitudinal strain and curvatures is summarized in the expression:

$$\mathbf{d}_i'(s) = [1 + \epsilon_3(s)] \cdot \boldsymbol{\kappa}(s) \times \mathbf{d}_i(s) \tag{2}$$

for $i = 1, 2, 3$ and where $\cdot'$ represents the spatial derivative with respect to $s$.

The steady-state governing equations of the configuration are obtained from Newtonian mechanics as

$$\mathbf{n}'(s) + \mathbf{f}(s) = \mathbf{0} \tag{3}$$

and

$$\mathbf{m}'(s) + \mathbf{l}(s) + \mathbf{r}'(s) \times \mathbf{n}(s) = \mathbf{0} \tag{4}$$

where $\mathbf{n}$ and $\mathbf{m}$ are the internal forces and moments, respectively. In (3) and (4), $\mathbf{f}$ and $\mathbf{l}$ represent the external forces and moments, respectively, acting on the arm per unit length. These external forces and moments include contributions from gravity, buoyancy, and tendon loads. All terms expressed in (3) and (4) are defined with respect to the inertial coordinate system attached to the base of the arm.

Assuming linear-elastic behavior, the constitutive relations for the Cosserat arm model are expressed as

$$\begin{bmatrix} EI_1(s) & 0 & 0 \\ 0 & EI_2(s) & 0 \\ 0 & 0 & GI_3(s) \end{bmatrix} \cdot \boldsymbol{\kappa}(s) = \mathbf{m}(s) \tag{5}$$

and

$$\rho A(s) \cdot \epsilon_3(s) = n_3(s) \tag{6}$$

where $E$, $G$, $\rho$, and $A$ are the Young's modulus, the modulus of rigidity, the mass density, and the cross-sectional area of the arm, respectively. The terms $I_i$ represent the moment of inertia of the cross section with respect to axis $\mathbf{d}_i$, and $n_3$ is the component of $\mathbf{n}$ perpendicular to the cross section.

Equations (2)–(6) constitute the kinematic Cosserat model of the soft arms utilized in forward kinematics calculations and dataset generation. Explicit expressions of the external forces $\mathbf{f}$ and moments $\mathbf{l}$ can be obtained based on tendon geometry. A two-stage numerical scheme outlined in [13] is employed for computing the forward kinematics solution where the first stage involves solving for the linear strain $\epsilon_3$ and curvatures $\boldsymbol{\kappa}$ for a given set of tendon forces from tip to base using (3)–(6). In the second stage, the position $\mathbf{r}$ and orientation $\mathbf{R}$ are calculated for the previously computed configuration space parameters using (2).

#### 4.1.2 Dataset Generation

To train the KineFormer, a forward kinematics dataset was generated from the analytical model of the soft arm described in the previous section. The input to this forward model is the forces in the tendons, and the output is the coordinates of the discrete points of the material curve in 3D space. Uniformly randomly distributed tension forces in the range of $0 \leq T_i \leq 10$ N are generated for each tendon $i$. The corresponding tip position of the arm computed from this forward kinematic model is recorded to form a labeled training dataset. Data is generated in the form of episodes, where each episode consists of 200 steps. Each episode is initialized at the nominal state of the soft arm with all tendon forces equal to zero, and the arm is in the undeformed configuration with the tip at the rest position. The statistics of the training dataset are presented in Table 1.

### *4.2 SOFA Arm*

SOFA (Simulation Open Framework Architecture) [14] is an open-source framework designed for physics simulation. The software is designed in particular for modeling and simulation of soft and deformable materials. It allows creating and customizing simulation models for various applications such as medical and soft

**Table 1** Statistics of the training dataset. All results are presented in mm

| Metric | $x$-axis | $y$-axis | $z$-axis | Dist. from base | Dist. from rest |
|---|---|---|---|---|---|
| Min | 95.1 | −306.2 | −305.2 | 184.4 | 0.0 |
| Max | 610.0 | 305.4 | 305.5 | 610.0 | 545.4 |
| Mean | 308.1 | 0.0 | 0.0 | 391.2 | 375.5 |
| Std. dev. | 114.1 | 159.6 | 159.4 | 75.9 | 118.5 |

**Table 2** Statistics of the SOFA model dataset. All results are presented in mm

| Metric | $x$-axis | $y$-axis | $z$-axis | Dist. from base | Dist. from rest |
|---|---|---|---|---|---|
| Min | −107.8 | −118.4 | 60.9 | 71.5 | 0.0 |
| Max | 114.8 | 105.7 | 195.0 | 195.0 | 140.3 |
| Mean | −3.5 | −1.9 | 167.2 | 176.4 | 53.4 |
| Std. dev. | 37.5 | 33.5 | 21.7 | 14.8 | 36.1 |

robotics. SOFA simulation is based on the FEM used to model the nonlinear deformation of the soft structure, actuators, and surroundings. To further evaluate the effectiveness of the KineFormer algorithm for controlling robot arm movements, a series of simulations were conducted using the SOFA physics simulation platform.

### 4.2.1 Finite Element Arm Model

The simulations were performed on a soft robot arm model actuated by eight cables and implemented in SOFA using FEM. The model of the arm was first introduced in [15]. This soft robotic arm called Trunk is actuated by eight cables; four cables actuate the first section of the arm and four go through the whole arm to the tip. The length of the arm is 195 mm.

### 4.2.2 Dataset Generation

The soft arm model in SOFA physics simulation is used to generate the dataset to train the KineFormer for the second experiment. The input to the SOFA model is the cable displacements with a range of $0 \leq C_i \leq 70$ mm for the cables that span the entire arm and $0 \leq C_i \leq 40$ mm for the cables that control the first section of the arm. Similar to the first experiment, the dataset is generated in the form of episodes with the starting state being the nominal state (all cable displacements are equal to zero). The statistics of the generated dataset are presented in Table 2.

## 5 Results and Discussion

As is discussed in Sect. 3, a separate KineFormer model is trained using the episodes of the datasets generated from each simulation. A total of 80% of the generated datasets were used for training the models, and the remaining 20% were utilized as the test datasets. The transformer architecture operates on fixed sequence lengths. Therefore, a sequence length of $N = 25$ elements is chosen for training the KineFormer. The embedding size is set to 128, and the architecture of the KineFormer consists of 12-layer decoder with an 8-head self-attention layer is utilized.

The desired actuation values $\mathbf{A}_d(n)$ are masked with zeros during training. The input element for step $n$ is defined as $\boldsymbol{\tau}(n) = [\mathbf{s}(n), \mathbf{r}_d(n), \mathbf{0}]$. The loss function

$$\mathcal{L} = \frac{1}{N}\sum_{n=0}^{N-1}[\mathbf{A}_d(n) - \mathbf{A}(n)]^2 \tag{7}$$

which represents the mean square error between the predicted actuation values $\mathbf{A}_d(n)$ and the analytically computed actual actuation values $\mathbf{A}(n)$ is employed for training the KineFormer.

A conventional FFNN was also trained on the dataset to serve as a baseline inverse kinematic model. The FFNN has two hidden layers of 256 neurons. The input layer is the desired tip position, and the output layer is defined as the predicted actuation values. This FFNN utilizes the same loss function given in (7). In contrast to the KineFormer which is trained on sequential data, the FFNN treats each step in the sequence independently from the rest of the dataset. Both the FFNN and KineFormer models are trained on a NVIDIA GeForce RTX 2060 GPU. The evolution of the loss functions for both models are provided in Fig. 3 for the two simulations. It was observed that at least 200 epochs were necessary to train the KineFormer, while 50 epochs was deemed sufficient for the training of the FFNN. This difference stems from the fact that the Kineformer possesses more parameters that require training compared to the FFNN.

A series of numerical simulations were conducted to test the performance of the KineFormer in predicting the necessary actuation values. In these simulations, desired tip positions were input to the network that outputs the required actuation values. The predicted actuation values were then fed into the analytic Cosserat model and the SOFA FEM model for the first and second simulations, respectively, to compute the corresponding tip positions. The first simulation represents a more static setup where the arm is commanded to reach random positions in 3D space,

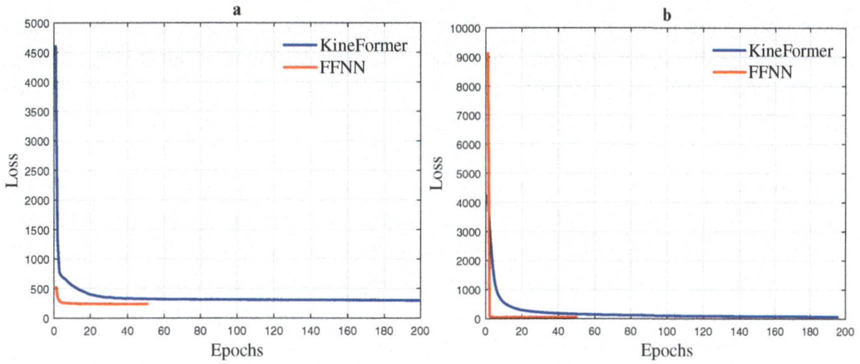


**Fig. 3** The training losses associated with the KineFormer and the FFNN models: (**a**) analytic Cosserat model and (**b**) the SOFA FEM simulations

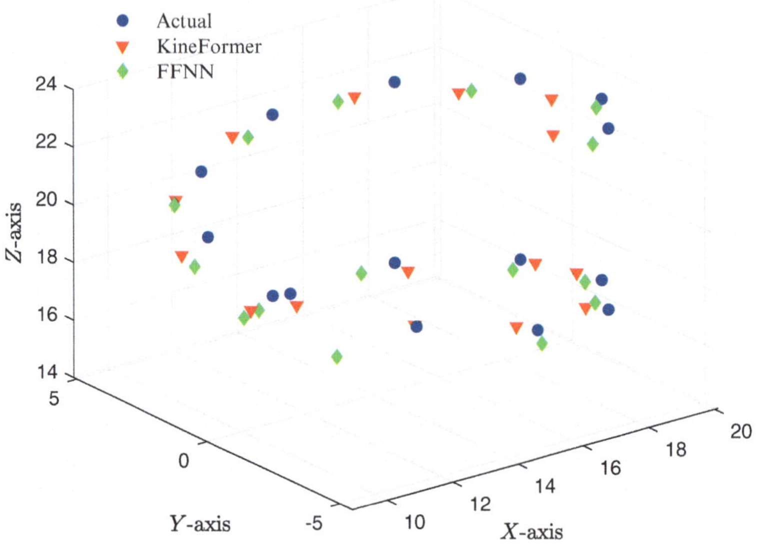


**Fig. 4** Typical tip positioning results provided by the KineFormer for the first experiment. The blue and red circles represent the desired and actual tip positions, respectively

while the second simulation is based on quasi-dynamic trajectory with sequential waypoints. Typical results obtained for the KineFormer in positioning the arm tip are depicted in Fig. 4 for the first experiment and Fig. 5 for the second experiment. Both results indicate that it is possible to position the arm at the desired point in 3D space using the KineFormer inverse kinematic model.

Furthermore, a benchmark analysis of the baseline FFNN model and the KineFormer network is performed. The performance of both networks is evaluated in terms of the mean absolute error and standard deviation in predicting the required actuation values (tendon forces for the first simulation or cable displacements for the second simulation) for a given desired tip position in 3D space. The results presented in Table 3 for the first simulation (analytic Cosserat model) indicate that both methods perform uniformly over the four cables. Results from the second (FEM) simulations that are given in Tables 4 and 5 indicate that the KineFormer has lower mean absolute error and standard deviation values compared to the FFNN model. The KineFormer produces more accurate and precise estimates of the actuation values in both simulations. In addition, the resulting tip positioning errors obtained from these predicted actuation values are also evaluated and compared for both simulations in Tables 6 and 7. In terms of positioning errors, the KineFormer outperforms the FFNN, producing a mean absolute positioning error of less than 5 mm in 3D space in contrast to the 10 mm provided by the latter.

The run-time performance of both methods are also evaluated by measuring the required computational time over several iterations. Accordingly, the FFNN and the KineFormer complete the computations in $115 \pm 6.14$ µs and $5920 \pm 189$ µs,

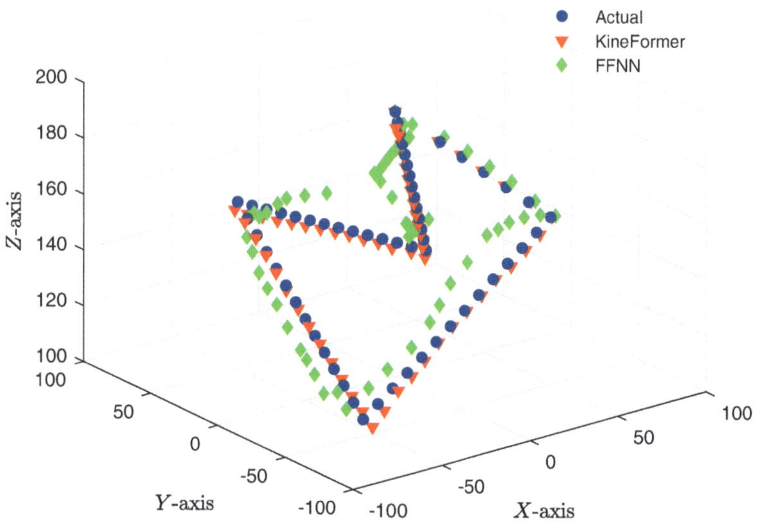


**Fig. 5** Typical tip positioning results provided by the KineFormer and the FFNN for the second experiment. The blue, red, and green points represent the actual and the predicted tip positions by KineFormer and FFNN, respectively

**Table 3** Performance results in terms of mean absolute error and standard deviation of the tendon force predictions for the FFNN and KineFormer networks. All results are in N

| Model | $T_1$ | $T_2$ | $T_3$ | $T_4$ |
|---|---|---|---|---|
| FFNN | $0.81 \pm 0.7$ | $0.82 \pm 0.7$ | $0.82 \pm 0.7$ | $0.81 \pm 0.7$ |
| KineFormer | $0.55 \pm 0.2$ | $0.55 \pm 0.2$ | $0.55 \pm 0.2$ | $0.55 \pm 0.2$ |

**Table 4** Performance results in terms of mean absolute error and standard deviation of the cable displacement predictions for the FFNN and KineFormer networks. All results are in mm

| Model | $C_1$ | $C_2$ | $C_3$ | $C_4$ |
|---|---|---|---|---|
| FFNN | $0.84 \pm 0.87$ | $0.84 \pm 0.82$ | $0.97 \pm 1.02$ | $0.95 \pm 1.05$ |
| KineFormer | $0.23 \pm 0.14$ | $0.38 \pm 0.17$ | $0.35 \pm 0.13$ | $0.28 \pm 0.15$ |

**Table 5** Performance results in terms of mean absolute error and standard deviation of the cable displacement predictions for the FFNN and KineFormer networks. All results are in mm

| Model | $C_5$ | $C_6$ | $C_7$ | $C_8$ |
|---|---|---|---|---|
| FFNN | $0.73 \pm 0.84$ | $0.74 \pm 0.96$ | $0.65 \pm 0.78$ | $0.68 \pm 0.79$ |
| KineFormer | $0.42 \pm 0.17$ | $0.36 \pm 0.24$ | $0.18 \pm 0.17$ | $0.30 \pm 0.20$ |

respectively. The longer computation time associated with the KineFormer is attributed to the higher complexity of the transformer network. However, these run-time results are deemed satisfactory for the real-time operation of the KineFormer in tele-operating soft arms.

**Table 6** Performance results in terms of mean absolute error and standard deviation for the tip positioning errors for the FFNN and KineFormer networks for the analytic Cosserat model. All results are in mm

| Model | $x$-axis | $y$-axis | $z$-axis |
|---|---|---|---|
| FFNN | $6.1 \pm 5.1$ | $4.8 \pm 3.7$ | $4.9 \pm 4.1$ |
| KineFormer | $3.5 \pm 3.1$ | $2.5 \pm 2.2$ | $2.8 \pm 2.5$ |

**Table 7** Performance results in terms of mean absolute error and standard deviation for the tip positioning errors for the FFNN and KineFormer networks for the SOFA FEM simulation. All results are in mm

| Model | $x$-axis | $y$-axis | $z$-axis |
|---|---|---|---|
| FFNN | $4.18 \pm 5.33$ | $4.36 \pm 4.98$ | $7.89 \pm 9.98$ |
| KineFormer | $2.38 \pm 4.27$ | $2.40 \pm 4.33$ | $4.21 \pm 4.03$ |

## 6 Conclusions

Machine learning methods represent a viable alternative for solving the difficult inverse modeling problem of soft robotic arms. In particular, soft robotic arms exhibit redundancy, whereby a given tip position may be achieved by multiple different configurations. Certain soft robotic manipulation activities such as robotic surgery or object manipulation in confined spaces may necessitate maintaining a specific arm shape. To address these issues the KineFormer network is proposed in the work. The KineFormer is based on the transformer architecture that has garnered significant attention in various non-robotic domains. The proposed KineFormer model is auto-regressive, in the sense that it takes past values into account before making decisions regarding future actions. This property can be exploited to ensure the consistency of the arm's shape while being controlled. Furthermore, for a given desired tip position in 3D space, it is shown that the KineFormer outperforms the baseline FFNN-based machine learning approach and can accurately predict the required actuation values. One limitation of the suggested methodology lies in its computational costs. Transformer-based structures like KineFormer necessitate considerable computational resources for both training and inference. Another constraint rests in the data prerequisites of transformer models; they require a substantial amount of data for efficient training. Nevertheless, the latter limitation can be overcome by pretraining a model using simulated data, prior to fine-tuning the model on actual data obtained from the arm. Although the scope of the current work is limited to static scenarios, the proposed model can readily be extended to handle sequential multi-waypoint or time-dependent (dynamic) problems for real-time shape and motion control of soft robotic arms.

## References


1. Zhang, Z., Bieze, T.M., Dequidt, J., Kruszewski, A., Duriez, C.: Visual servoing control of soft robots based on finite element model. In: IEEE/RSJ International Conference on Intelligent Robots and Systems (IROS), pp. 2895–2901 (2017)
2. Largilliere, F., Verona, V., Coevoet, E., Sanz-Lopez, M., Dequidt, J., Duriez, C.: Real-time control of soft-robots using asynchronous finite element modeling. In: IEEE International Conference on Robotics and Automation (ICRA), pp. 2550–2555 (2015)
3. Holsten, F., Engell-Norregard, M.P., Darkner, S., Erleben, K.: Data driven inverse kinematics of soft robots using local models. In: International Conference on Robotics and Automation (ICRA), pp. 6251–6257 (2019)
4. Gillespie, M.T., Best, C.M., Townsend, E.C., Wingate, D., Killpack, M.D.: Learning nonlinear dynamic models of soft robots for model predictive control with neural networks. In: IEEE International Conference on Soft Robotics (RoboSoft), pp. 39–45 (2018)
5. Bern, J.M., Schnider, Y., Banzet, P., Kumar, N., Coros, S.: Soft robot control with a learned differentiable model. In: 3rd IEEE International Conference on Soft Robotics (RoboSoft), pp. 417–423 (2020)
6. Kovandzic, M., Nikolić, V., Simonovic M., Ćiric, I., Al-Noori, A.: Soft robot positioning using artificial neural network. Facta Universitatis, Series: Automatic Control and Robotics, vol. 18(1), pp. 19–30 (2019)
7. Thuruthel, T.G., Falotico, E., Cianchetti, M., Laschi, C.: Learning global inverse kinematics solutions for a continuum robot. In: Proc. 21st CISM-IFToMM Symp. ROMANSY21-Robot Design, Dynamics and Control, vol. 569, pp. 47–54 (2016)
8. Vaswani, A., Shazeer, N., Parmar, N., Uszkoreit, J., Jones, L., Gomez, A. N., Kaiser, L., Polosukhhin, I.: Attention is all you need. Advances in Neural Information Processing Systems (NeurIPS), vol. 30 (2017)
9. Devlin, Chang, M.-W., Lee, K., Toutanova, K.: BERT: Pre-training of deep bidirectional transformers for language understanding. In: Proc. Conf. North American Chapter of the Association for Computational Linguistics (NAACL): Human Language Technologies, vol. 1, pp. 4171–4186, Minneapolis, MN (2019)
10. Radford, A., Narasimhan, K.: Improving language understanding by generative pre-training, unpublished. https://s3-us-west-2.amazonaws.com/openai-assets/research-covers/language-unsupervised/language_understanding_paper.pdf. Last access date: 30 Oct. 2022
11. Dosovitskiy, A., Beyer, L., Kolesnikov, A., Weissenborn, D., Zhai, X., Unterthiner, T., Dehghani, M., Minderer, M., Heigold, G., Gelly, S.: An image is worth 16x16 words: Transformers for image recognition at scale. ArXiv: 2010.11929 (2020)
12. Chen, L., Lu, K., Rajeswaran, A., Lee, K., Grover, A., Laskin, M., Abbeel, P., Srinivas, A., Mordatch, I.: Decision transformer: Reinforcement learning via sequence modeling. ArXiv: 2010.11929 (2021)
13. Renda, F., Cianchetti, M., Giorelli, M., Arienti, A., Laschi, C.: A 3D steady-state model of a tendon-driven continuum soft manipulator inspired by the octopus arm. Bioinspir. Biomim. **7**(2), 025006 (2012)
14. Faure, F., Duriez, C., Delingette, H., Allard, J., Gilles, B., Marchesseau, S., Talbot, H., Courtecuisse, H., Bousquet, G., Peterlik, I.: Sofa: A multi-model framework for interactive physical simulation. Studies in Mechanobiology Tissue Engineering and Biomaterials, pp. 283–321 (2012)
15. Coevoet, E., Escande, A., Duriez, C.: Optimization-based inverse model of soft robots with contact handling. IEEE Robot. Autom. Lett. **2**(3), 1413–1419 (2017)

# Part II
# Networks

# Multiple Node Localization in Cognitive Radio-Based Wireless Sensor Networks Using Grid Search

**Suzan Ureten**

## 1 Introduction

Robotics has been an active research field over the last decades in many different applications such as in industrial manufacturing, agriculture, and medicine. The advancements in new hardware and software development have given great impetus to this field of research. As a continuing part of the process, researchers have investigated collaborative aspects where a set of robots can work together to complete many different tasks autonomously, such as collaborative navigation. Navigation is a challenging issue for autonomous mobile robots due to many separate phases the robots must go through. These phases are perception, localization, cognition, and motion control. When the robots extract meaningful data from their employed environment in the perception phase, they estimate their current location by using the information from external sensors in the localization phase. The robots can then decide for their next step to take for a given target and modify their outputs accordingly, in the cognition and motion control phases, respectively.

Wireless sensor networks and mobile ad hoc networks have also been extensively studied by communication and networking researchers, where nodes or, namely, robotic sensors are considered either static or mobile. The studies made for robust communication in wireless networks have maintained significant development for secure and robust communication between the robots, where the communication channels between the robots can be highly dynamic in mobile robotic sensor networks. Robotic wireless sensor networks can be thought as autonomous networked multi-robot systems where robots exchange data among each other to maintain a required communication performance through collaborative work,

S. Ureten (✉)
Bahcesehir University, İstanbul, Türkiye
e-mail: suzan.ureten@bau.edu.tr

 
Ö. M. Gül et al. (eds.), *7th EAI International Conference on Robotic Sensor Networks*, EAI/Springer Innovations in Communication and Computing,
https://doi.org/10.1007/978-3-031-64495-5_4

learning, and adaptation. In wireless sensor networks where the communication takes place in unlicensed spectrum bands, high communication performances cannot be achieved due to overcrowding of the existing unlicensed spectrum. However, flexile radiotechnologies, such as cognitive radios that can access the spectrum when it is idle, have been the core subject of research in mitigating the insufficient spectrum problem. Cognitive radio (CR) networks must comply with spectrum regulations in the region in which they operate and prevent interference to primary licensees in their exclusion regions. Exclusion region of a primary user is the area within which the secondary license holders are not allowed to do transmission [12]. For predicting the exclusion region, location and transmission power information of the primary user must be known. For most cases, these information may not be readily available due to the nature of noncooperative wireless networks.

Estimating the location of the primary transmitter in a noncooperative radio network is more difficult than the standard localization problem due to the lack of cooperation or feedback from the emitters. In the case of having one emitting node in a sensor network, the location of the emitting node can be estimated by using trilateration technique where only three received power measurements are required. Alternatively, if there are a large number of sensors measuring the received power due to the emitting node, the least squares estimate technique can be used. However, in the case of having multiple emitting nodes in a noncooperative network, the observed power at each sensor is, in fact, the addition of received power values due to all emitters in the network. Especially, in noncooperative robotic networks, there is no feedback between the robotic nodes, and the power contribution of each emitting node in a receiving node is not known. For such noncooperative network scenarios, straightforward solutions to the problem of localizing the emitters cannot be carried out easily [9]. This is because isolating each transmitting node requires complex signal processing algorithms in the robotic sensors. Particularly, in dynamic networks, white areas do not last long and the estimation time needs to be short enough for a cognitive node to detect and use white areas in the spectrum. Therefore, localization algorithms which require less time are preferred in CR-based wireless networks.

The literature on multi-node localization thrives on different algorithms for accurate location estimation. In [6], the average power and location estimation problem of two emitting sources are examined. Similarly, in [7, 8], and [13], the global optimization approach, the quasi expectation maximization (EM) technique, and an iterative multi-primary user localization algorithm that non-collaboratively predicts locations based on k-means clustering are proposed for localizing multiple emitters under log-normal shadowing. In this paper, we consider applying the maximum-likelihood estimation (MLE) method based on grid search algorithm for the localization of two emitting nodes with an equal transmit power value by utilizing the received power measurements obtained at multiple sensors as the MLE offers an attractive approach for the localization problems. The MLE method is attractive because it is asymptotically efficient and unbiased and prior location information is not required [11].

One challenge of the grid search algorithm is that probability calculations at each grid location require an analytical expression to sums of log-normal random variables (RVs) representing a fraction of the received power values for each pair of emitting nodes. However, there is no analytical expression for the sum of log-normal variables. There are various suggestions for finding approximate analytical expressions in the literature, and in most of them, approximating the sum of log-normal RVs with another log-normal RV is preferred. Fenton-Wilkinson [4], Schwartz-Yeh [10], and Beaulieu-Xie [1] methods are based on this approach.

Among the methods mentioned above, we consider using Fenton-Wilkinson's approach because Fenton's approach is a computationally efficient method and it can be applicable to our model assumptions as the actual received power values at the sensors are non-iid (nonindependent and identically distributed) log-normal RVs; i.e., they have different mean power values and different dB spread values for each transmitter-receiver pair. Most of the existing approximation methods in the literature are only applicable to iid random variables and do not work for our model. On the other hand, in terms of prediction accuracy, Fenton's approach fails when the shadowing is high. Therefore, we propose a new sum of log-normal approximation based on curve fitting technique which results in much better estimation accuracies at high shadowing spread values as well as at low spread values for different network case scenarios. The highly accurate sum of log normal approach we propose is our novel contribution.

# 2 Problem Statement

## 2.1 *Network Model*

The considered CR-based wireless sensor network is described as follows:

- $N_s$ secondary mobile sensors are placed at known but arbitrary locations in a region of interest (ROI). These nodes are robotic sensors that measure the received power values due to the emitting nodes that exist in the area.
- There are $N_t$ primary emitting nodes transmitting simultaneously with equal transmit power at a given frequency and bandwidth.
- Bandwidth and frequency of the emitting nodes are known or readily detectable parameters.
- The secondary robotic sensors collect power and report their power measurements along with their location information to a fusion center, where the information is further processed to estimate the location of the transmitting nodes in the network.
- Individual sensor measurements are assumed to be perfect.

### 2.2 Signal Model

- The received power measured at each sensor has a distance dependent path loss model.
- The average received power at at the $j$th receiver $j = 1, 2, \ldots, N_s$ from the $i$th primary emitter, $i = 1, 2, \ldots, N_t$ is given by

$$\mu_{ij} = \frac{\beta P_t}{(d_{ij})^\alpha} \tag{1}$$

- $P_t$ is the power of one emitting node and it is equal to the power of the other emitting node.
- $\beta$ is a constant representing the carrier frequency and antenna characteristics, $\beta = (\frac{\lambda}{4\pi})^2$, $\lambda$ is the wavelength.
- $\alpha$ is the path-loss exponent.
- $d_{ij}$ is the distance between sensor $j$ and the $i$th primary emitting node.
- Each sensor experiences log-normal shadowing. The unknown measured power from the $i$th emitting node to the $j$th sensor is given by

$$r_{ij} = \mu_{ij} 10^{\frac{W}{10}} \tag{2}$$

  where $W \sim \mathcal{N}(0, \sigma^2)$ is the gain/loss in dB due to shadowing.
- $\sigma$ is the shadowing spread. It is known and takes the same value for each transmitter-receiver pair.
- The received power at the $j$th sensor, $r_j$, is the sum of the received powers values due to all emitting nodes:

$$r_j = \sum_{i=1}^{N_t} r_{ij} \tag{3}$$

- As $r_{ij}$ is a log-normally distributed RV, $r_j$ has a probability density function (pdf) that is a sum of log-normal distributions which doesn't have any closed form expression.

## 3 A Sum of Log-Normal Approximation from Simulated Data

There are several proposed methods for searching a pdf to sum of log-normals in literature. Considering our model assumptions and the localization algorithm we use, there are two major criteria for selecting the appropriate approximation among the earlier studies mentioned in Sect. 1. Firstly, the pdf approximation needs to be applicable to non-iid log-normal random variables because the mean value of the power, that is, the total received power at each sensor due to the emitters, will differ and will have the standard deviation, that is, the dB spread for each transmitter-receiver pair. Secondly, the approximation technique needs to be computationally efficient as the grid-search algorithm inherently requires high number of calculations due to the need to scan every unique emitter-pair location configuration at the grid centers. Consequently, due to its simplicity and applicability to our model assumptions, we choose Fenton's approximation for having an analytical expression for the sum of log-normals pdf. This approximation is shown in (6) and assumes that the sum of log-normal RV is another log-normal RV that has a pdf with mean $\mu_j$ and variance $\Omega_j^2$ given as follows:

$$f(r_j|\mu_j, \Omega_j^2) = \frac{1}{\sqrt{2\pi} r_j \Omega_j} e^{-\frac{(\ln(r_j)-\mu_j)^2}{2\Omega_j^2}} \tag{4}$$

where

$$\mu_j = \ln\Big(\sum_{i=1}^{N_t} \mu_{ij} + \epsilon^2 \frac{\sigma^2}{2} - \frac{\Omega_j^2}{2}\Big) \tag{5}$$

$$\Omega_j^2 = \ln\Big((e^{\epsilon^2\sigma^2} - 1)\frac{\Sigma_{i=1}^{N_t}\mu_{ij}^2}{(\Sigma_{i=1}^{N_t}\mu_{ij})^2} + 1\Big) \tag{6}$$

and $\epsilon = \frac{\ln 10}{10}$.

In Fenton-Wilkinson method, the first two central moments of random variables are matched so that the mean and the standard deviation of the log-normal sum distribution can be computed. In this method, when the standard deviation of the summands is less than 4 dB, the first two central moments can be matched. However, when the standard deviation of the summands are more than 4 dB, the method breaks down in computation of first and the second moments [3, 10]. This situation limits the use of Fenton's approach only to cases where the shadowing spread values are less than 4 dB. For mitigating this problem, we propose a novel method that also approximates the log-normal sum with another log-normal distribution where its parameters are estimated from a set of simulated data as described below. Although the problem statement given in Sect. 2 is a general description for localization of multiple emitters, let us consider two emitters ($N_t = 2$) case for simplicity without loss of generality.

Suppose $Y_1$ and $Y_2$ are log-normally distributed RVs with different means of $\mu_{1j}$ and $\mu_{2j}$, but same standard deviations of $\sigma$, where $\mu_{1j}$ and $\mu_{2j}$ are the mean received power values at the $j$th sensor due to the first and the second emitter, respectively. Then, $Y$, which is the summation of these log-normal RVs $Y = Y_1 + Y_2$, will have a sum of log-normal distribution. We model the sum of log-normal distribution with another log-normal distribution with the new parameters $\mu_j$, $\Omega_j$ which are functions of $\mu_{1j}$, $\mu_{2j}$, and $\sigma$:

$$Y_1 \sim LN(\mu_{1j}, \sigma) \tag{7}$$

$$Y_2 \sim LN(\mu_{2j}, \sigma) \tag{8}$$

$$Y \sim SLN(\mu_j, \Omega_j) \tag{9}$$

where $\mu_j = f_1(\mu_{1j}, \mu_{2j}, \sigma)$ and $\Omega_j = f_2(\mu_{1j}, \mu_{2j}, \sigma)$. Here, the symbols $LN(\cdot)$ and $SLN(\cdot)$ denote the log-normal and sum of two log-normal distributions, respectively. For generating the simulated data, first two separate log-normal RVs $Y_1$ and $Y_2$ are simulated and then added. The resultant RV denoted by $Y$ composes the simulated data. A log-normal distribution is fitted to the simulated data $Y$, and the parameters of the fitted log-normal distribution are estimated by using the maximum likelihood technique [2].

The simulation study performed in this paper has the following setup parameters: The mean received power values $\mu_{1j}$ and $\mu_{2j}$ at the $j$th sensor due to two emitting nodes are bounded in between $-100$ dBm and 0 dBm. Using a resolution of 1 dB, $101 \times 101 = 10,201$ different combinations of mean received power values $(\mu_{1j}, \mu_{2j})$ are generated. The generated mean received power values are then stored in a look-up table (LUT). One million random numbers are generated from two distinct log-normal distributions with the parameters of $(\mu_{1j}, \sigma)$ and $(\mu_{2j}, \sigma)$ for each $(\mu_{1j}, \mu_{2j})$. The summation of these simulated numbers is further used as random numbers that have sum of log-normal distribution. These random numbers are fitted to a log-normal distribution and the new parameters $\mu_j$ and $\Omega_j$ are estimated. These parameters are then used to calculate the likelihood of observing the simulated data which is an essential step in an emitter localization algorithm. As another part of our simulation study, our proposed approach is compared with Fenton's approach in terms of emitter location estimation accuracies via the grid search method described below.

## 4 Localization Using Grid Search Algorithm

We consider a square ROI in which two emitting nodes are located. This ROI is divided into a number of grids. The center of the grids is set as test locations for the two emitting nodes. We start the search from the first test locations as if the first emitting node is located at the center of the first grid, whereas the second emitting node is located at the remaining grid centers, successively. We continue our search by changing the test location of the first emitting node to the grid center next to it and locating the second emitting node at the rest of the grid centers, successively. This searching iteration is repeated for all emitter-pair test location configurations. Repeated emitter-pair test location configurations are omitted from the iterations as they are unnecessarily increasing the number of computations. The likelihood of observing the received power measurements at all sensors for each unique emitter-pair configuration are calculated by

$$L = \prod_{j=1}^{N_s} \frac{1}{\sqrt{2\pi} r_j \Omega_j} e^{-\frac{(\ln(r_j)-\mu_j)^2}{2\Omega_j^2}} \tag{10}$$

This likelihood expression is derived by taking the product of the pdf expression given in (4) for all sensors under the assumption of independent log-normal shadowing for each transmission. Among all test locations, the emitter-pair configuration which maximizes the likelihood value is an estimate of emitters' locations.

A matching of the estimated locations to each emitting node is not required because their transmit power values are equal; therefore, the estimated location of one emitting node would belong to any of the two emitting nodes.

## 5 Performance Evaluations

The localization performance of the proposed sum of log-normal approach based on distribution fitting is compared with Fenton-Wilkinson's approach by applying the grid search algorithm to four different network case scenarios with the given network assumptions and constraints. These scenarios are small area dense network, small area sparse network, large area dense network, and large area sparse network cases.

For all scenarios, we consider a microcellular network that has a cluster size of 3 ($S = 3$) and a cell size of 100 ($C = 100$ m). The reuse distance $R$, which is the required distance between base stations operating at the same frequency, can then be calculated as $R = C\sqrt{3S} = 300$ m. This preference brings a constraint on the emitter geometry in the ROI such that the minimum distance between the emitting nodes has to be 300 m. Otherwise, the two emitting nodes would interfere each other, which reflects the physical fact that primary co-channel emitters would interfere when they are located too close to each other.

The distance between the sensors and the emitters is assumed to be at least twice the reference so that the log-distance propagation model yields realistic results in the far-field region [5]. The value of this distance is taken as 10 m or 100 m, depending on the selected network scenario. Sensors are also placed at least 10 m or 100 m apart from each other in the ROI for different network case scenarios. The transmit frequency is set to 900 MHz which is a typical frequency of GSM, and the propagation exponent is set to 2.5. Received power values measurements at sensors are in the range of $-102$ dBm and $-36$ dBm. Therefore, the minimum received power is calculated to be higher than the receiver sensitivity with a typical value of $-112$ dBm and less than the maximum input power rating with a typical value of 10 dBm. For small area network scenarios, the ROI is taken as 1 km by 1 km square area. The minimum distance between the emitters and the sensors and the minimum distance within the sensors are taken as 10 m. The power of the emitting node is set to 1 W and the grid size is set to 50 m. For large area network scenarios, the ROI is taken as 3 km by 3 km square area. The minimum distance between the emitters and the sensors and the minimum distance within the sensors are also taken as 10 m for dense networks; however, for sparse networks, the minimum distance between the emitters and the sensors and the minimum distance within the sensors are taken as 100 m. In dense networks, 100 sensors are deployed, whereas in sparse networks, 20 sensors are deployed to collect the RSS measurements. For large network scenarios, the power of the emitting node is set to 10 W, and the grid size is set to 150 m.

Throughout the simulations, parameters $(\mu_j, \Omega_j)$ are extracted from the LUT for each emitter-pair location configuration and each sensor's location. However, For Fenton's approach, these parameters are calculated by using Eqs. (5) and (6). For both approaches, likelihood values are then calculated as shown in (10). The overall localization error is obtained by averaging the first and the second emitter localization errors, $e_1$ and $e_2$, respectively. 500 simulations are run for different dB spread values ranging from 1 dB to 10 dB. For four different network case scenarios, emitter localization errors with respect to dB values are plotted in Figs. 1, 2, 3, 4.

As can be seen from Figs. 1, 2, 3, 4, the localization error is less in dense network scenarios where the number of sensors is larger. When the area is larger, the localization error increases. In all these case scenarios, our proposed pdf approximation approach provides smaller localization error values compared to Fenton-Wilkinson's approach for all shadowing spread values.

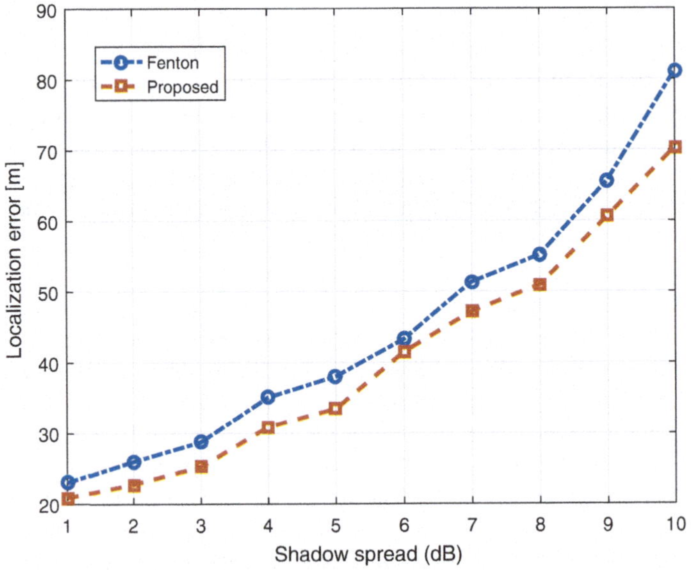


**Fig. 1** Performance comparison for small area dense network scenario

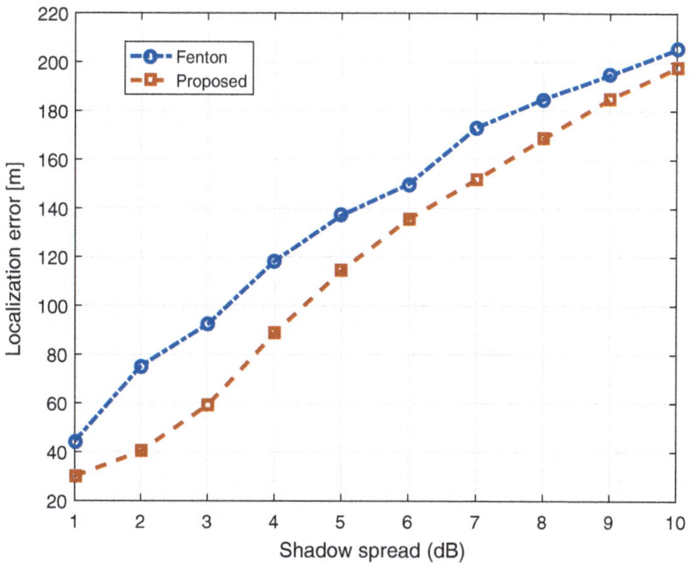


**Fig. 2** Performance comparison for small area sparse network scenario

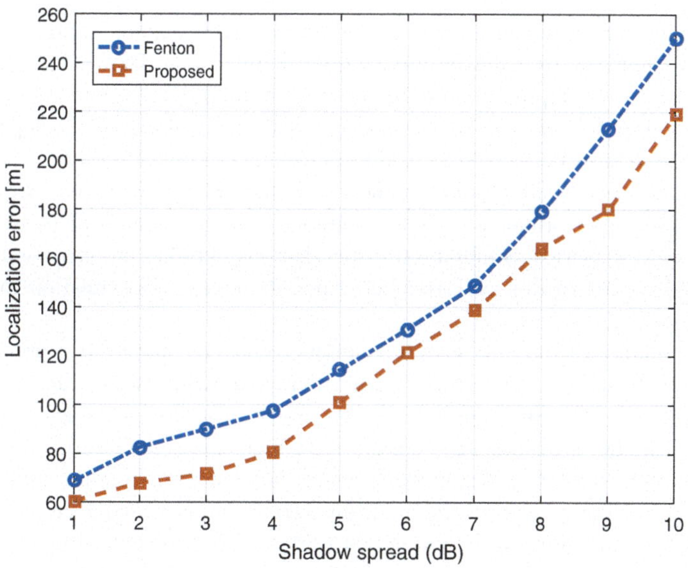


**Fig. 3** Performance comparison for large area dense network scenario

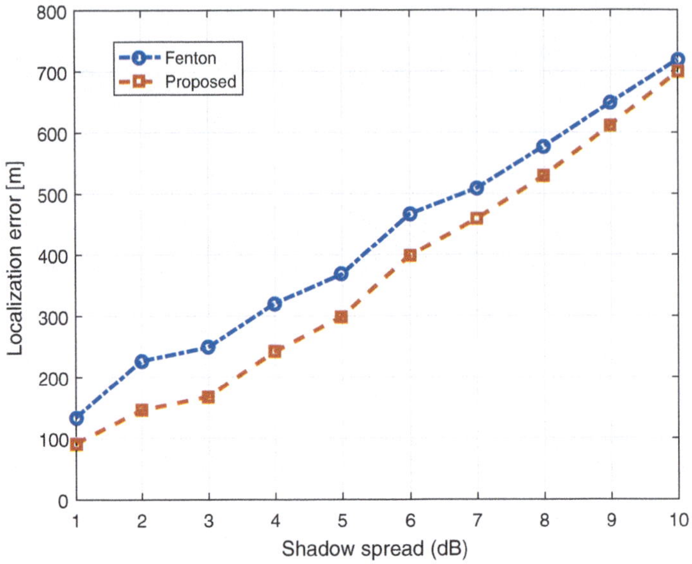


**Fig. 4** Performance comparison for large area sparse network scenario

## 6 Conclusion

In this study, we presented a novel probability density function approximation to sum of log-normals by using a curve fitting method to a data set generated via simulations. In our approximation, we fit a log-normal distribution to the data set which is composed of sum of log-normal random numbers. We compared our approximation with Fenton-Wilkinson's approximation that is known to be simple and adequately performing for low shadowing values. For comparison, we evaluated the localization performance of both techniques via grid search method for four different case scenarios. Simulation results demonstrate that the proposed sum of log-normal approximation performs better than Fenton's approximation for all dB spread values.

The limitation of the grid search technique is mainly due to its computational complexity as it is a bruce-forced algorithm and requires to scan every grid location in the ROI. In order to increase the localization accuracies with this technique, smaller grid sizes can be preferred which comes with a cost of increased computational complexity. Additionally, when the number of emitters are increased in the ROI, the complexity of the algorithm would be increased even more as the number of computations of the likelihood values at all grid centers depends on the number of emitting nodes. For future studies, our proposed approximation can be applied in less complex localization algorithms for which a of sum of log-normal approximation is required.

## References

1. Beaulieu, N.C., Xie, Q.: An optimal lognormal approximation to lognormal sum distributions. IEEE Trans. Veh. Technol. **53**(2), 479–489 (2004)
2. Bowman, A.W., Azzalini, A.: Applied Smoothing Techniques for Data Analysis: The Kernel Approach with S-Plus Illustrations, vol. 18. OUP Oxford (1997)
3. Cardieri, P., Rappaport, T.S.: Statistical analysis of co-channel interference in wireless communications systems. Wirel. Commun. Mobile Comput. **1**(1), 111–121 (2001)
4. Fenton, L.: The sum of log-normal probability distributions in scatter transmission systems. IRE Trans. Commun. Syst. **8**(1), 57–67 (1960)
5. Goldsmith, A.: Wireless Communications. Cambridge University Press, Cambridge (2005)
6. Miliou, N., Moustakas, A., Polydoros, A.: Interference source localization and transmit power estimation under log-normal shadowing (2011)
7. Nelson, J.K., Gupta, M.R., Almodovar, J.E., Mortensen, W.H.: A quasi em method for estimating multiple transmitter locations. IEEE Signal Process. Lett. **16**(5), 354–357 (2009)
8. Nelson, J.K., Hazen, M., Gupta, M.R.: Global optimization for multiple transmitter localization, pp. 1–7 (2006)
9. Raman, C., Kalyanam, J., Seskar, I., Mandayam, N.: Distributed spatio-temporal spectrum sensing: An experimental study. In: IEEE Forty-First Asilomar Conference on Signals, Systems and Computers, pp. 2063–2067 (2007)
10. Schwartz, S.C., Yeh, Y.: On the distribution function and moments of power sums with log-normal components. Bell Syst. Tech. J. **61**(7), 1441–1462 (1982)
11. Sengijpta, S.K.: Fundamentals of statistical signal processing: Estimation theory. Technometrics **37**(4), 465–466 (1995)
12. Vu, M., Devroye, N., Tarokh, V.: On the primary exclusive region of cognitive networks. IEEE Trans. Wireless Commun. **8**(7), 3380–3385 (2009)
13. Zhiqiang, X., Wei, L., Qiang, H.: An iterative multiple primary users localization algorithm based on clustering. In: 6th International ICST Conference on Communications and Networking in China (CHINACOM), pp. 241–244. IEEE (2011)

# Systematic Review of Network Slicing Resource Management in 5G

**Ahmed Mostafa Elbaz, Heba Kamal Aslan, and Islam Tharwat Abdel-Halim**

## 1 Introduction

The evolution of mobile networks over the past four decades has brought about substantial transformations. The initial two iterations of mobile communication technology encompassed voice and text functionalities, whereas the third generation (3G) introduced the concept of broadband connectivity, characterized by data transmission speeds reaching several hundred kilobits per second (Kbps). The data rates of 4G networks have experienced a significant increase, reaching speeds of tens of megabits per second. The objective of 5G extends beyond the mere enhancement of data rates. The advent of 5G technology facilitates a shift from a singular access service, namely broadband connectivity, to a more expansive array of services encompassing mission-critical applications, extensive Internet-of-Things capabilities, and improved mobile broadband performance. [1].

As shown in Fig. 1, 5G, like other cellular networks, consists of two components. The first component is the radio access network (RAN), which is called new radio (NR) in 5G, and it manages the radio spectrum. The second component is the Mobile Core, which is also known as the next-generation core (NG-Core) in 5G, and it is in charge of tracking where users move and connecting data and voice services to the Internet [2].

A. M. Elbaz
School of Information Technology and Computer Science, Nile University, Giza, Egypt
e-mail: a.elbaz@nu.edu.eg

H. K. Aslan · I. T. Abdel-Halim (✉)
School of Information Technology and Computer Science, Nile University, Giza, Egypt

Center for Informatics Science (CIS), Nile University, Sheikh Zayed, Egypt
e-mail: haslan@nu.edu.eg; ihalim@nu.edu.eg



Ö. M. Gül et al. (eds.), *7th EAI International Conference on Robotic Sensor Networks*, EAI/Springer Innovations in Communication and Computing,
https://doi.org/10.1007/978-3-031-64495-5_5

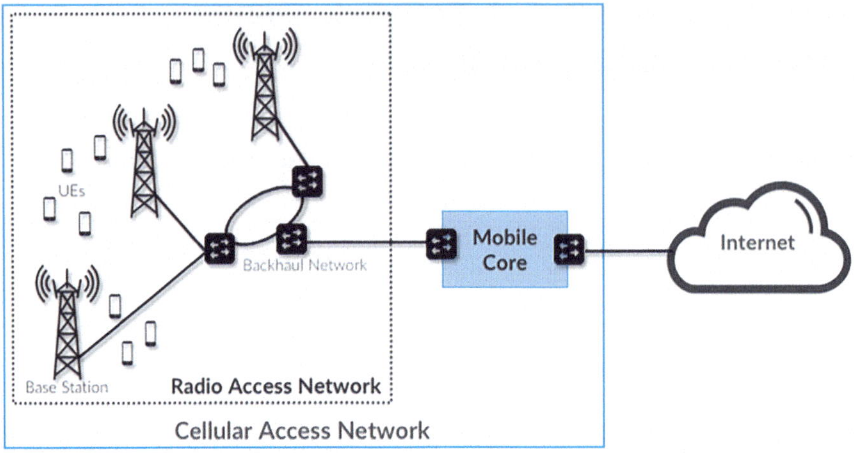


**Fig. 1** Cellular networks' main components [2]

Besides, 5G services are divided into three types [3]: There are three key requirements for next-generation wireless communication systems. Firstly, enhanced mobile broadband (eMBB) necessitates high data rates. Secondly, massive machine-type communication (mMTC) demands ultra-high density. Lastly, ultra-reliable low-latency communication (uRLLC) necessitates both ultra-high availability and ultra-low latency. In order to maintain cost-effectiveness, it is necessary to optimize resources to accommodate diverse demands. Network slicing can be employed to accomplish this objective through the utilization of software-defined networking (SDN) and network functions virtualization (NFV) technologies as fundamental elements within the architecture of the 5G network. This approach facilitates the division of the network into distinct slices, each of which can be allocated for specific use cases [4]. With network slicing, 5G networks can be dynamically adjusted to the different needs of industries and services. To get to this level of dynamism, a lot of research was done to figure out how to allocate and arrange the different resources [5].

The main contribution of this chapter is to propose a taxonomy of the currently used techniques for network slicing resource management based on the main 5G network components (RAN, Core, or E2E) through a systematic review of experiences reported in the literature starting from 2019. Future research directions to improve resource management will also be discussed.

The rest of the chapter is organized as follows: In Sect. 2, the criteria for selecting the papers used in the systematic review are explained. Sections 3 and 4 discuss the current trends and future research directions of network slicing resource management in 5G. Finally, the conclusion is presented in Sect. 5.

## 2 Methodology

Before discussing the paper's selection criteria, we need to explain why we selected a systematic literature review in this survey paper from the different types of surveys. The main characteristics of a systematic review are a clear research question with inclusion and exclusion criteria; a strict and systematic search of the literature; definitive appraisal of included studies; data management and extraction; analysis and interpretation of results; and a report for publication. Based on these characteristics, the systematic review's main aim is to reduce bias at all stages of the review process. Although less robust study designs can contain bias, systematic reviews are not immune to introducing bias.

We have two research questions we are trying to answer. The first question is "What are the current trends or research directions in resource management in 5G network slicing?" The second is "What are the open research areas in resource management in 5G network slicing?" First, we searched for questions that discuss the usage of artificial intelligence in network slicing resource management but we found a lot of research related to resource management that does not use AI. So, we decided to go with more generic questions and check the current trends and open research areas regardless of whether AI is used or not.

We focused on our search of the top publishers' databases only to have consolidated results of each selected paper; so we searched for papers from the following databases: IEEE Xplore (http://ieeexplore.ieee.org/Xplore/home.jsp), Springer (https://link.springer.com), ScienceDirect (http://www.sciencedirect.com), and MDPI (MDPI—Publisher of Open Access Journals).

The search string used to narrow the search criteria includes the terms "5G" and "Network Slicing" in the paper title and contains the "resource management" keyword in the document, noting that the exact search string differs from one database to another as the search criteria are not identical between all of them. The initial list includes 146 papers, noting that the used search string of each database and the output number of papers per database are provided in Table 1.

**Table 1** Search string for each database

| | IEEE Xplore | Springer | ScienceDirect | MDPI |
|---|---|---|---|---|
| Search string | Search term "5G" in Document Title AND "Network Slicing" in Document Title AND "resource management" in ALL Metadata | Where the title contains "network slicing" 5G—with the exact phrase "resource management" | Title: 5G—network slicing, find articles with these terms: resource management | 5G: Title AND "network slicing": Title AND resource management: all fields |
| Number of output papers | 90 | 27 | 23 | 6 |

**Table 2** Number of papers after applying inclusion/exclusion criteria

| | IEEE Xplore | Springer | ScienceDirect | MDPI | Total |
|---|---|---|---|---|---|
| Papers published between 2019 and 2022 | 66 | 26 | 20 | 6 | 118 |
| Articles/journals only | 23 | 6 | 19 | 6 | 54 |
| Survey paper exclusion | 21 | 6 | 16 | 6 | 49 |
| Quality assessment | 16 | 2 | 6 | 3 | 27 |

The next step of the performed systematic review is defining the inclusion and exclusion criteria. We defined two inclusion criteria. The first focuses on articles and journals, which as a result means conferences will be out of this survey scope. The second includes papers published between 2019 and 2022. Also, two exclusion criteria were defined. The first excludes survey papers and the second performs a quality assessment and excluding disqualified papers. The quality assessment was based on the scores given by the three researchers working on this paper. Each researcher provided a score from 0 to 3 (3 is the most qualified) for each paper, then the average score was calculated and the paper with a score of less than 2 was considered disqualified. The provided score was based on three aspects: whether the paper is relevant to the research question, a clear problem statement, and whether there are clear findings. After applying the above inclusion/exclusion criteria, we got 27 papers to perform our analysis on. The inclusion/exclusion criteria and the total number of papers after each inclusion/exclusion step are shown in Table 2.

## 3 Current Trends

In this section, we will discuss the first research question: What are the current trends or research directions in resource management in 5G network slicing? As discussed in the introduction section, 5G network consists of two main components: RAN and core network. Hence, in order to address this research question, the papers are classified into four categories: (i) end-to-end (radio and core included) network slicing resource management; (ii) RAN side slicing resource management; (iii) core side slicing resource management; and (iv) other components like transport layer, wearable devices, etc.

Based on the above classification, we can say that out of the 27 papers we worked on, there were 6 end-to-end papers, 10 RAN papers, 4 core papers, and 7 classified as others. In the following sections, we will present the efforts exerted in each class separately.

## 3.1 End-to-End (E2E)

Many papers focused on E2E resource allocation in 5G network slices. In [4], the authors presented a cross-layer control framework that leverages software-defined networking (SDN) and software-defined radio (SDR) techniques. The integration of the software-defined networking (SDN) control plane and the software-defined radio (SDR) data plane into a cohesive architecture will significantly improve both domains. The primary objective is to optimize network performance in radio environments characterized by the presence of diverse radio access technologies.

In [5], a Deep Q-Networks (DQN) algorithm that fits multiservice and multislice was proposed to dynamically allocate resources. The algorithm considers both radio access network slices and core network slices. The main goal is to maximize the access users, and when compared to the optimal allocation of the access side there is a 9% increase in the average access rate for delay-constrained slices.

In [6], the authors proposed a genetic algorithm that can solve network function allocation problems for multiservice 5G networks with reduced complexity and execution time. This approach considers both RAN and core sides and deploys network functions on a distributed environment based on the service requests. The main objective of the chapter is to provide a fast-and-easy deployment of the network functions.

In [7], a real-time data-driven approach was proposed to realize joint slice reconfiguration from 5G core to 5G RAN. A slice reconfiguration algorithm is used to load balance the slices in 5G core and 5G RAN. The main aim of the chapter is to reduce the cost of slice status updates and enhance slice reconfiguration performance.

As indicated in [8], a proposed queuing model encompasses three sequential queuing model subsystems, namely cloud radio access network (C-RAN), mobile edge computing (MEC), and cloud data center. This model aims to estimate the quality of service (QoS) parameters for 5G networks and determine the requisite number of C-RAN and MEC cores under varying 5G traffic conditions.

In [9], the authors directed their attention toward the development of an end-to-end framework specifically designed for 5G vehicular ad hoc networks. The utilization of a genetic algorithm (GA) enables the system to adapt dynamically to the ever-changing network load conditions and attain the optimal allocation of bandwidth slices, all while meeting the desired performance objectives.

## 3.2 Core

Managing cloud resources in general and mobile core resources is crucial for better system utilization and hence better performance. In this context, Lagrangian relaxation (LR) method optimization-based approach was used for allocating VMs to servers effectively and efficiently [10]. The main aim of the LR method is to maximize system utilization while achieving the quality of services and resource requirements.

In [11], the authors proposed the algorithm RT-CSP, which is a two-stage provisioning algorithm for the 5G core network slices. RT-CSP comprises a heuristic algorithm for slice node provisioning and a slice link provisioning algorithm based on k-shortest paths. The primary aim of this study is to effectively allocate 5G core network slices in order to optimize resource utilization within the 5G core network infrastructure, thereby leading to an increase in revenue for the infrastructure provider (InP).

In [12], the authors focused on the efficient utilization of the network edge by developing a network slicing model based on a cluster of fog nodes (FNs) that are managed by an edge controller (EC). The EC decides for each service request which FN in the cluster can handle it or just reject it and return it back to the cloud. The problem was formulated as an infinite-horizon Markov decision process (MDP) and was solved using a deep reinforcement learning (DRL) solution that gradually learns the optimal slicing policy.

In [13], the authors presented the network-slicing management and orchestration (NS-MANO) solution, which incorporates the novel enablers for cloud slicing (NECOS) platform. This integration aims to overcome the constraints of existing cloud computing infrastructures in meeting the requirements for novel services and facilitating end-to-end cloud network slicing. The objective of this study is to employ the NECOS slicing concept to establish the cloud-to-things continuum.

### *3.3 RAN*

First, we will present the papers that tried to solve problems related to resource allocation. In [14–16], the authors focused on providing solutions for efficient RAN resource allocation taking the required quality of service (QoS) per slice into consideration. In [14], the authors proposed a genetic algorithm (GA) intelligent latency-aware resource allocation scheme (GI-LARE), which targets to enhance resource allocation for multitier, multitenant heterogeneous environments. Besides, the authors in [15] proposed a deep reinforcement learning (DRL)-based network slicing model to find the resource allocation policy that maximizes throughput while satisfying the QoS requirements. DRL was used to let the gNB (agent) check the delay requirements, channel state, and data rate constraints (states), and then maximize the overall throughput (reward) by assigning each resource block (RB) to a certain service (action). In [16], a new framework, namely, new radio flexibility (NRflex), was introduced and mapped to the O-RAN architecture. The main objective of the proposed framework is to maximize user equipment (UE) scheduling and optimize PRB allocation while fulfilling the required QoS. An algorithm was proposed to run near time to dynamically compute the bandwidth part (PWP) size dedicated to a certain RAN slice. Also, another multiplexing and scheduling algorithm was introduced to run at the gNB to periodically decide the active BWP to be used by each UE and the amount of PRB assigned to it.

In [17], the objective of the authors was to optimize the performance of the weighted-sum transmission rate while adhering to a constraint on delay. This was achieved by proposing an appropriate algorithm for resource allocation in the uplink direction. The problem of simultaneously optimizing subchannel allocation and power control was formulated as a Markov decision process (MDP) problem with an infinite time horizon and average-reward constraints. The derivation of the optimal control policy is initially accomplished through the utilization of the value iteration algorithm. Next, the Q-factor for subchannel allocation is broken down into the Q-factor for each individual slice. The Q-factor and Lagrangian multipliers are updated through the utilization of online stochastic learning algorithms.

In [18], the authors directed their attention toward time division duplex (TDD) networks in the context of 5G, with a specific emphasis on improving the efficiency of resource utilization. This work introduces two mechanisms for resource reconfiguration: (i) the integration of slice resource allocation and test-driven development (TDD) framework selection, and (ii) the allocation of slices with independent distributed uplink/downlink (UL/DL) ratio adjustment.

Besides, in [19], an intelligent resource scheduling strategy (iRSS) for 5G RAN slicing was introduced to improve resource-multiplexing gain among slices while meeting specific QoS. The integration of deep learning (DL) and reinforcement learning (RL) forms the basis of the intelligent RSS (iRSS) system. Deep learning (DL) is commonly employed for the purpose of conducting resource allocation on a large time scale. On the other hand, reinforcement learning (RL) is utilized for online resource scheduling, which addresses network dynamics on a smaller time scale. This includes handling unexpected network states and mitigating inaccuracies in prediction.

In [20], the authors introduced a Markov model for RAN slicing that can define diverse radio resource management (RRM) strategies in multiservice and multitenant 5G scenarios. The model defines both GBR and non-GBR and also captures 5G scenarios with different radio propagation conditions through a new statistical model.

In [21, 22], the authors worked on RAN slicing resource orchestration. In [21], a discrete-time Markov chain (DTMC) model was introduced to capture the dynamics of the aggregate slice traffic demand and the wireless channel variations to simultaneously provide latency and throughput guarantees in a multitenancy environment. In [22], machine learning (ML) techniques were used for resource orchestration. The model includes four building blocks that contain demand classification, optimum slicing ratios prediction, admission control, and adaptive resource management. Regression trees (RTs) are used for demand classification. The decision-maker block is composed of two sub-blocks: a forecast-aware slicer using ML-based regression and an admission controller that grants or denies resource requests based on the current or forecasted load. Finally, deep reinforcement learning (DRL) is implemented for the adaptive resource manager building block.

## 3.4 Others

In this section, we will introduce the efforts exerted in various domains other than RAN and core like transport network, 5G wearable networks, and D2D-based 5G-heterogeneous cellular network (5G-HCN).

In [24, 25], the authors worked on transport network slicing. In [24], the authors aimed to achieve greater flexibility and optimize resource utilization in the 5G optical mobile fronthaul (MFH) by using an integrated network resource management (INRM) scheme that mutually considers MFH bandwidth allocation and cloud computing offloading. Traffic offloading helps fulfill the various slices' heterogeneous needs and ensures fair and efficient resource sharing among slices. Also, the main focus of the proposed work in [25] was to optimize the backhaul transport network latency and improve load balance. Hence, a branch-and-bound algorithm is used to provide the optimal physical path that the slice traffic follows.

In addition, in [26], an admission control mechanism for network slicing was used to maximize network infrastructure providers' revenue while reducing costs. Similarly, in [27], authors proposed a machine learning policy-based admission mechanism that ensures efficient resource utilization and fairness to the service providers. Also, in [28], the authors aimed to efficiently manage resources of network slicing in 5G wearable networks. The service cognitive engine is used to analyze the service features and the resource cognitive engine monitors the usage of resources. Finally, a global cognitive engine is used to achieve service-aware resource allocation.

In [29], the objective of this study was to optimize the average network spectrum efficiency (SE) and throughput in a device-to-device (D2D)-based 5G network while ensuring that the overall system performance remains unaffected. The paper proposes a distributed resource allocation strategy utilizing an edge computing (DRA-EC) approach to identify a suboptimal solution. The distributed augmented Lagrange method involves the individual edge routers located at base stations (BS) solving the problem locally. Subsequently, a consensus algorithm is employed to obtain the global solution by utilizing the local estimates from the edge routers.

In [30], a machine learning algorithm was introduced with the aim of effectively provisioning 5G network slices by classifying specific network slices such as "eMBB, mMTC, and URLLC" for individual devices. The model comprises three primary stages: the process of gathering information or data for research purposes. One variation of optimal weighted feature extraction (OWFE) is being discussed. The weight function was optimized through the hybridization of two meta-heuristic algorithms, namely glowworm swarm optimization and deer hunting optimization algorithm (DHOA). This hybrid model was named glowworm swarm-based DHOA (GS-DHOA). The classification technique of slicing is employed by a hybrid classifier that combines neural networks and deep belief.

# 4 Future Research Directions

In this section, we will address the second research question What are the open research areas in resource management in 5G network slicing? We can divide the main research directions into three parts. The first part is applying the different models to large-scale networks. The second part is using ML to get more efficient resource allocation/orchestration models. The final part is related to working in a multitenant environment instead of a single one, so let us discuss each part individually.

## *4.1 Large-Scale Networks*

In some papers, the results were obtained by applying the different algorithms to simulators or small-scale live networks. For future work, it is suggested to test these algorithms on larger-scale networks.

In [9], the simulation showed that the proposed slicing framework is able to optimize resources in 5G vehicular ad hoc networks. The future work is to apply the framework of complex networks, including RAN and CN, to compete for resources during transmission, buffering, and computing. In [16], the NRflex framework that is used for efficient resource allocation was implemented in OpenAirInterface (OAI) 5G to test it in real deployment.

Also, in [17], the proposed algorithm improved the user transmission rate on a RAN coverage area with a radius of 100 m through efficient resource allocation for uplink communication. Hence, it is of vital importance to analyze the results on multiple base stations in the proposed RAN slicing networks to make the optimization more challenging.

In [23], the network slicing strategies were tested when a new slice request arrives, which was not taken into consideration in the paper's analysis. In [24], a high-level simulation showed that traffic offloading helps in supporting the heterogeneous needs of various 5G fronthaul slices while ensuring efficient and fair resource sharing among slices. Future work will focus on evaluating the model in a large-scale network scenario instead of the high-level simulation.

In [26], admission policies that were tried on the MATLAB simulator were examined and proposed for integration in real systems, while in [27], the neural networks (NN)-based policy-based admission mechanism showed resource efficiency and fairness to service providers when applied to MATLAB simulator and the authors plan to extend the solution to include the case with different service classes and more complex network.

## 4.2 Machine Learning–Based Algorithms

Many papers suggested using ML algorithms in future research to get more efficient resource management systems. For example, in [4], a cross-layer control framework was introduced, which exploits SDN and SDR techniques. The framework optimizes the network performance in dense radio environments. Deep reinforcement learning techniques can be used in the future in order to further improve the network's performance.

Also, in [6], the basic configurations obtained using the genetic algorithm (GA) were used to train a neural network that provides real-time solutions for network function deployment based on instantaneous data.

Besides, the authors in [14] proposed a genetic algorithm intelligent latency-aware resource allocation scheme (GI-LARE) for efficient radio resource management in a multitenant, multitier 5G RAN slice network. They also mentioned using a generative adversarial network (GAN) for more efficient resource management as they think GANs will make resource allocation more dynamic compared to the genetic algorithm.

In [13], the network slicing management and orchestration (NS-MANO) was refined by devising machine learning techniques that employ a proactive strategy through a predictive load approach. In [29], a new resource allocation scheme based on leveraging machine learning and deep reinforcement learning techniques was suggested to further improve the resource allocation compared to the currently used distributed resource allocation strategy with an edge computing (DRA-EC) approach.

## 4.3 Multitenancy

The concept of multitenancy in 5G is to enable mobile virtual network operators to share the resources from infrastructure providers by requesting and leasing resources dynamically on demand. In [10], an optimization-based approach (LR) was used to efficiently utilize computing resources, which maximizes system utilization, subject to the quality of services, and resource requirements. As a next step, resource sharing between multiple service providers can be explored, which necessitates considering issues such as the sharing economy between multiple network service providers.

In [20], a Markov model for RAN slicing resource management was introduced, and in the future, it is desired to extend the resource allocation procedure to have joint optimization of the performance of different tenants by distributing the unused PRBs of one tenant among other tenants.

## 5 Conclusion

This study undertook a comprehensive analysis of resource management for network slicing in 5G networks through a systematic review. The present study delved into the dynamic nature of the 5G landscape, encompassing its varied service prerequisites, and emphasized the pivotal significance of network slicing in fulfilling these prerequisites. The systematic review conducted in this study was predicated upon an exhaustive exploration of esteemed databases. Furthermore, stringent criteria were employed to determine the eligibility of papers, thus ensuring the high caliber of the selected literature.

The selected papers were classified into four primary categories: end-to-end (E2E), core, RAN, and additional domains encompassing transport networks, wearables, and D2D-based 5G-heterogeneous cellular networks (5G-HCN). The various categories provided valuable insights pertaining to resource management techniques, effectively demonstrating the diverse and intricate nature of 5G network slicing.

A notable trend that was observed in the field of end-to-end resource management was the incorporation of novel methodologies such as Deep Q-Networks (DQN), genetic algorithms, and real-time data-driven solutions in order to enhance the efficiency of resource allocation. The strategies employed in this study were centered around the optimization of access rates, improvement of network performance, and attainment of low latency for various services.

By and large, the management of network slicing resources in the context of 5G is a domain that is characterized by its dynamic nature and continuous development. It holds considerable promise in enhancing network performance and optimizing resource allocation. Future research should further investigate novel methodologies, expand the scope to encompass larger networks, and harness the potential of machine learning to effectively address the multifaceted requirements of the 5G environment. The growing significance of 5G networks across multiple industries necessitates the imperative role of resource management optimization in facilitating improved services and experiences.

## References

1. Chapter 1: Introduction — 5G Mobile Networks: A Systems Approach Version
2. Chapter 3: Basic Architecture — 5G Mobile Networks: A Systems Approach Version 1.1-dev documentation
3. Subedi, P., Alsadoon, A., Prasad, P.W.C., et al.: Network slicing: a next generation 5G perspective. J. Wireless. Com. Network. **2021**, 102 (2021). https://doi.org/10.1186/s13638-021-01983-7
4. Barmpounakis, S., Maroulis, N., Papadakis, M., Tsiatsios, G., Soukaras, D., Alonistioti, N.: Network slicing – enabled RAN management for 5G: Cross layer control based on SDN and SDR. Comput. Netw. **166**, 106987., ISSN 1389-1286 (2020). https://doi.org/10.1016/j.comnet.2019.106987

5. Li, T., Zhu, X., Liu, X.: An end-to-end network slicing algorithm based on deep Q-learning for 5G network. IEEE Access. **8**, 122229–122240 (2020). https://doi.org/10.1109/ACCESS.2020.3006502
6. Shinde, S.S., Marabissi, D., Tarchi, D.: A network operator-biased approach for multi-service network function placement in a 5G network slicing architecture. Comput. Netw. **201**, 108598., ISSN 1389-1286 (2021). https://doi.org/10.1016/j.comnet.2021.108598
7. Yang, H., et al.: Data-driven network slicing from core to RAN for 5G broadcasting services. IEEE Trans. Broadcast. **67**(1), 23–32 (2021). https://doi.org/10.1109/TBC.2020.3031742
8. AlQahtani, S.A., Alhomiqani, W.A.: A multi-stage analysis of network slicing architecture for 5G mobile networks. Telecommun. Syst. **73**, 205–221 (2020). https://doi.org/10.1007/s11235-019-00607-2
9. Khan, A.A., Abolhasan, M., Ni, W., Lipman, J., Jamalipour, A.: An End-to-End (E2E) network slicing framework for 5G vehicular Ad-Hoc networks. IEEE Trans. Veh. Technol. **70**(7), 7103–7112 (2021). https://doi.org/10.1109/TVT.2021.3084735
10. Hsiao, C.-H., Lin, F.Y.-S., Fang, E.S.-H., Chen, Y.-F., Wen, Y.-F., Huang, Y., Su, Y.-C., Wu, Y.-S., Kuo, H.-Y.: Optimization-based resource management algorithms with considerations of client satisfaction and high availability in elastic 5G network slices. Sensors. **2021**, 21 (1882). https://doi.org/10.3390/s21051882
11. Li, X., Guo, C., Xu, J., Gupta, L., Jain, R.: Towards efficiently provisioning 5G core network slice based on resource and topology attributes. Appl. Sci. **9**, 4361 (2019). https://doi.org/10.3390/app9204361
12. Nassar, A., Yilmaz, Y.: Deep reinforcement learning for adaptive network slicing in 5G for intelligent vehicular systems and smart cities. IEEE Internet Things J. **9**(1), 222–235 (2022). https://doi.org/10.1109/JIOT.2021.3091674
13. Maciel, D.B., Neto, E.P., Costa, K.B., et al.: Cloud-network slicing MANO towards an efficient IoT-cloud continuum. J. Grid Comput. **19**, 48 (2021). https://doi.org/10.1007/s10723-021-09588-6
14. Oladejo, S.O., Falowo, O.E.: Latency-aware dynamic resource allocation scheme for multi-tier 5G network: a network slicing-multitenancy scenario. IEEE Access. **8**, 74834–74852 (2020). https://doi.org/10.1109/ACCESS.2020.2988710
15. Suh, K., Kim, S., Ahn, Y., Kim, S., Ju, H., Shim, B.: Deep reinforcement learning-based network slicing for beyond 5G. IEEE Access. **10**, 7384–7395 (2022). https://doi.org/10.1109/ACCESS.2022.3141789
16. Boutiba, K., Ksentini, A., Brik, B., Challal, Y., Balla, A.: NRflex: enforcing network slicing in 5G new radio. Comput. Commun. **181**, 284–292., ISSN 0140-3664 (2022). https://doi.org/10.1016/j.comcom.2021.09.034
17. Song, F., Li, J., Ma, C., Zhang, Y., Shi, L., Jayakody, D.N.K.: Dynamic virtual resource allocation for 5G and beyond network slicing. IEEE Open J. Veh. Technol. **1**, 215–226 (2020). https://doi.org/10.1109/OJVT.2020.2990072
18. Shrivastava, R., Samdanis, K., Sciancalepore, V.: Towards service-oriented soft spectrum slicing for 5G TDD networks. J. Netw. Comput. Appl. **137**, 78–90., ISSN 1084-8045 (2019). https://doi.org/10.1016/j.jnca.2019.01.009
19. Yan, M., Feng, G., Zhou, J., Sun, Y., Liang, Y.-C.: Intelligent resource scheduling for 5G radio access network slicing. IEEE Trans. Veh. Technol. **68**(8), 7691–7703 (2019). https://doi.org/10.1109/TVT.2019.2922668
20. Vilà, I., Pérez-Romero, J., Sallent, O., Umbert, A.: Characterization of radio access network slicing scenarios with 5G QoS provisioning. IEEE Access. **8**, 51414–51430 (2020). https://doi.org/10.1109/ACCESS.2020.2980685
21. Zanzi, L., Sciancalepore, V., Garcia-Saavedra, A., Schotten, H.D., Costa-Pérez, X.: LACO: a latency-driven network slicing orchestration in beyond-5G networks. IEEE Trans. Wireless Commun. **20**(1), 667–682 (2021). https://doi.org/10.1109/TWC.2020.3027963
22. Salhab, N., Langar, R., Rahim, R.: 5G network slices resource orchestration using Machine Learning techniques. Comput. Netw. **188**, 107829 (2021., ISSN 1389-1286). https://doi.org/10.1016/j.comnet.2021.107829

23. muñoz, P., Adamuz-Hinojosa, ñ., Navarro-Ortiz, J., Sallent, O., Pérez-Romero, J.: Radio access network slicing strategies at spectrum planning level in 5G and beyond. IEEE Access. **8**, 79604–79618 (2020). https://doi.org/10.1109/ACCESS.2020.2990802
24. Song, C., et al.: Hierarchical edge cloud enabling network slicing for 5G optical fronthaul. J. Opt. Commun. Netw. **11**(4), B60–B70 (2019)
25. Li, W., et al.: Latency-optimal virtual network functions resource allocation for 5g backhaul transport network slicing. Appl. Sci. **9**(4), 701 (2019)
26. Vincenzi, M., Lopez-Aguilera, E., Garcia-Villegas, E.: Maximizing infrastructure providers' revenue through network slicing in 5G. IEEE Access. **7**, 128283–128297 (2019)
27. Vincenzi, M., Lopez-Aguilera, E., Garcia-Villegas, E.: Timely admission control for network slicing in 5G with machine learning. IEEE Access. **9**, 127595–127610 (2021)
28. Hao, Y., et al.: Data-driven resource management in a 5G wearable network using network slicing technology. IEEE Sens. J. **19**(19), 8379–8386 (2018)
29. Nadeem, L., et al.: Efficient resource allocation using distributed edge computing in D2D based 5G-HCN with network slicing. IEEE Access. **9**, 134148–134162 (2021)
30. Abidi, M.H., et al.: Optimal 5G network slicing using machine learning and deep learning concepts. Comput. Stand. Interfaces. **76**, 103518 (2021)

# Part III
# Computer Vision

# Comparatively Studying Modern Optimizers Capability for Fitting Vision Transformers

**Abdullah Nazhat Abdullah and Tarkan Aydin**

## 1 Introduction

Transformer architectures [1] are highly capable deep learning models that were introduced initially in the context of language modeling and relied mainly on the attention mechanism for their processing. Encouraged by the success of Transformer architectures in natural language processing (NLP), computer vision research also started to adopt such designs, and the Vision Transformer [2] is a prime example of employing attention in image processing tasks. It addresses the problem of processing the image input in a similar manner to language by applying the Transformer layers directly to embedded image patches without the need for a convolutional neural network (CNN) to extract features, thus opening up the prospects of unified and generalized architectures for both Vision and NLP domains. Such kind of efforts leads to highly generalized and capable models that operate on inputs from various modalities, such as Text, Images, Videos, and Audio providing with the same model a set of outputs for a multitude of tasks that correspond to each modality, such as text completion, sentiment analysis, image classification, image segmentation, and speech to text transcription [3]. In addition, such models can provide outputs with mixed modalities such as visual question answering, open vocabulary detection, and open vocabulary segmentation [3]. Due to the high performance and generalization of these models, it is commonly referred to such designs in the literature as foundational models [4]. Other Transformer architectures were introduced for computer vision tasks that incorporate different structures than the Vision Transformer (ViT). MLPMixer [5] and ConvMixer [6] in particular

A. N. Abdullah (✉) · T. Aydin
Bahcesehir University, İstanbul, Türkiye
e-mail: nazhat.abdullah@bahcesehir.edu.tr; tarkan.aydin@bahcesehir.edu.tr

 
Ö. M. Gül et al. (eds.), *7th EAI International Conference on Robotic Sensor Networks*, EAI/Springer Innovations in Communication and Computing,
https://doi.org/10.1007/978-3-031-64495-5_6

are Transformer architectures aimed from the inception to address image-based problems.

Optimization is a central part of training a deep learning model; it is the process by which the weights of the model are updated based on the error signals calculated from the gradients of the specified loss function of that model. Choosing the proper optimization scheme is vital for obtaining a model that is free from overfitting issues and delivers the best possible performance on testing samples. Selecting the best optimization algorithm is crucial in the process of designing advanced computer vision systems based on Transformer architectures due to the fact that the optimization scheme guides the learning of the deep learning model. Transformers lack a comprehensive and comparative analysis of optimizer effects that have been available to CNN architectures, which have had many studies assessing how various optimizers affect their performance. This comparative study measured the impact of choosing a specific optimizer when utilizing Transformer-based architectures in computer vision tasks. A collection of optimizers was identified belonging to different optimization paradigms, including a novel and newly introduced optimizer that was generated by an algorithm; the optimizers were used to train multiple image classifiers as a quantifiable assessment process, measuring the suitability and stability of each optimizer for fitting Vision Transformer models.

## 2 Related Work

This paper follows the findings of assessing the performance of different optimization algorithms on deep learning systems presented by Choi et al. [7]. Prior to that work, Wilson et al. [8] investigated a classification problem in which the Adaptive Moment (Adam) optimizer converged to provably worse solutions than the standard gradient methods and also presented experiments in which Adam produced worse validation accuracy than stochastic gradient descent (SGD). Schneider et al. [9] introduced a benchmark for neural network optimizers and reported that there was no single best optimizer across the workloads they considered. Dogo et al. [10] compared the performance of seven optimizers on Natural Images dataset, Cats and Dogs dataset, and Fashion MNIST dataset. Each optimizer was benchmarked on accuracy metric and convergence time. The evaluation used a simple CNN architecture with three convolutional layers and one dense layer with 64 neurons. Prilianti et al. [11] comparatively evaluated the performance of seven optimizers on the digital plant dataset. To assess each optimizer, three CNN architectures were used: the first was a shallow network with only one convolutional layer and without any dense layers; the second CNN architecture used was the LeNet architecture; and the third CNN architecture was the AlexNet. Jangid and Srivastava [12] compared three optimizers on handwritten Devanagari characters. The optimizers tested were Adam, Adamax, and Root Mean Square Propagation (RMSProp). The chosen architecture was a CNN with three convolutional layers and one dense layer with 1000 neurons. Swastika et al. [13] chose to compare three optimizers on vehicle type classification and utilized Adam, Adadelta, and SGD optimizers. The comparison adopted three CNN architectures to evaluate each optimizer with a shallow network: LeNet and MiniVGGNet. Many examples can be found in the literature for utilizing the Transformer architecture for computer vision tasks. Deng et al. [14] combined CNN and ViT for the task of classifying high-resolution remote sensing images. Yu et al. [15] proposed an enhanced ViT for fundus image classification, while Xue et al. [16] proposed deep hierarchical ViT for hyperspectral and light detection and ranging data classification. Tanzi et al. [17] utilized ViT in order to obtain a classification model of the femur fracture. Beal et al. [18] used a supervised pretrained ViT for object detection. Li et al. [19] studied vanilla ViT on object detection in the context of masked image modeling pretrained representations. Hatamizadeh et al. [20] introduced UNetFormer to address the task of medical image segmentation. Yuan et al. [21] designed TransAnomaly for the detection of anomalies in videos.

# 3 Methodology

## *3.1 Selected Transformer Architectures*

To obtain a good coverage of the analysis, three different Transformer architectures were selected that span a diverse and important section of available computer vision designs: ViT, MLPMixer, and ConvMixer.

The ViT architecture [1] applies attention-based blocks with the minimum possible modifications. Input images are divided into patches, and after linear embedding of these patches, they are processed by the Transformer architecture in the same way that word tokens are processed in NLP. It was shown experimentally that the ViT architecture, despite not incorporating inductive biases present in CNN architectures such as translation equivariance and locality, attains favorable results given that the training was performed on sufficient amounts of data.

The MLPMixer [2] is a simple Transformer architecture that drops the use of both convolutions and attention and adopts a design that uses multilayer perceptrons. These MLPs are used to mix the information in each block, either per the spatial locations or per the feature channels. This design utilizes basic operations such as matrix multiplications and scalar nonlinearities to transform the linearly projected patches with two phases of operation: the first is channel-mixing to perform information communication between the different representation channels (dimensions); and the second is token-mixing to communicate information between spatial locations of the token representations.

The ConvMixer [3] is a Transformer architecture that uses convolutions to mix the information presented in the image patches while maintaining the size and resolution of these patches by omitting the downsampling operation that usually interleaves traditional CNNs. Such a simple architecture is shown to be competitive with ResNets, which have a similar number of parameters and also have better small-size dataset performance compared to recent Transformers such as ViT and MLPMixer. This design highlights the impact of converting the input images to patches that are then projected into representations and also shows that uniform block architectures have an advantage over the traditional CNN design.

## 3.2 Selected Optimizers

For the task of comparatively assessing optimizer impact on the performance of the selected models, a diverse set of optimizers is chosen that represent different schemes and techniques, including the most recent additions to the literature. The designated optimizers are Adaptive Gradient (Adagrad), Adam, Adam with decoupled weight decay (AdamW), and Evolved Sign Momentum (LION).

Adagrad [22] is an optimizer that utilizes online bound calculations to improve the convergence rates and generalization of the optimization process in an adaptive manner, outperforming the preceding nonadaptive methods. The method uses a diagonal approximation of the subgradients matrix computed within the runtime of the optimizer. Below is the pseudocode for the Adagrad optimizer:

**Algorithm 1** Adagrad Optimizer

**Require:** Learning rate $\eta$, Small constant $\epsilon$, Initial parameters $\theta_0$

Initialize $r_0 = 0$
**while** stopping criterion not met **do**
    Compute gradient $g_t$ using current mini-batch
    $r_t = r_{t-1} + g_t^2$
    Update parameters: $\theta_t = \theta_{t-1} - \frac{\eta}{\sqrt{r_t} + \epsilon} \cdot g_t$
**end while**

Adam [23] is a first-order gradient approach that is efficient and provides low memory requirements. Adam computes adaptive learning rates for the target parameters to be optimized by estimating the first and second moments of the gradients, hence the name adaptive moment estimation. Several advantages are obtained from this approach to optimization, including scale-invariant magnitudes of parameter updates with respect to scaling the gradients, the ability to work with sparse gradients, and implicitly annealing the update step size. Below is the pseudocode for the Adam optimizer:

**Algorithm 2** Adam Optimizer

**Require:** Learning rate $\eta$, Exponential decay rates $\beta_1$, $\beta_2$, Small constant $\epsilon$, Initial parameters $\theta_0$

Initialize $m_0 = 0$, $v_0 = 0$, $t = 0$
**while** stopping criterion not met **do**
  $t = t + 1$
  Compute gradient $g_t$ using current mini-batch
  $m_t = \beta_1 \cdot m_{t-1} + (1 - \beta_1) \cdot g_t$
  $v_t = \beta_2 \cdot v_{t-1} + (1 - \beta_2) \cdot g_t^2$
  $\hat{m}_t = \frac{m_t}{1-\beta_1^t}$
  $\hat{v}_t = \frac{v_t}{1-\beta_2^t}$
  Update parameters: $\theta_t = \theta_{t-1} - \frac{\eta}{\sqrt{\hat{v}_t + \epsilon}} \cdot \hat{m}_t$
**end while**

AdamW [24] is an improved version of the Adam optimizer with a decoupled weight decay procedure, allowing for better generalization while maintaining the favorable adaptive qualities of the original Adam. Below is the pseudocode for the AdamW optimizer:

**Algorithm 3** AdamW Optimizer

**Require:** Learning rate $\eta$, Exponential decay rates $\beta_1$, $\beta_2$, Small constant $\epsilon$, Weight decay $\lambda$, Initial parameters $\theta_0$

Initialize $m_0 = 0$, $v_0 = 0$, $t = 0$
**while** stopping criterion not met **do**
  $t = t + 1$
  Compute gradient $g_t$ using current mini-batch
  $m_t = \beta_1 \cdot m_{t-1} + (1 - \beta_1) \cdot g_t$
  $v_t = \beta_2 \cdot v_{t-1} + (1 - \beta_2) \cdot g_t^2$
  $\hat{m}_t = \frac{m_t}{1-\beta_1^t}$
  $\hat{v}_t = \frac{v_t}{1-\beta_2^t}$
  Update parameters: $\theta_t = \theta_{t-1} - \frac{\eta}{\sqrt{\hat{v}_t + \epsilon}} \cdot (\hat{m}_t + \lambda \cdot \theta_{t-1})$
**end while**

LION [25] is a novel and recent optimization algorithm that was extracted from a process of algorithm discovery by program search . The search process utilized evolutionary search with warm-start and restart, abstract execution, funnel selection, and program simplification to address the issues of the infinite and sparse program space and select algorithms that offer good generalization and performance. This algorithm that was discovered by search is different from previous adaptive optimizers in the feature of tracking only the gradient momentum and using the sign operation to obtain the updates. Below is the pseudocode for the LION optimizer:

**Algorithm 4** Lion Optimizer

**Require:** Parameters $\beta_1, \beta_2, \lambda, \eta$, Objective function $f$
```
Initialize θ_0, m_0 ← 0
while θ_t not converged do
    Compute gradient
    g_t = ∇_θ f(θ_{t-1})
    Update intermediate parameters:
    c_t = β_1 m_{t-1} + (1 − β_1) g_t
    Update model parameters:
    θ_t = θ_{t-1} − η_t (sign(c_t) + λθ_{t-1})
    Update EMA of g_t:
    m_t = β_2 m_{t-1} + (1 − β_2) g_t
end while
return θ_t
```

## 4 Experiments and Results

Two datasets have been selected for experimentation: CIFAR-10 and CIFAR-100. The CIFAR-10 dataset [26] consists of 60000 color images in 32 by 32 resolution provided for 10 classes, with 6000 images per class. There are 50000 training images and 10000 test images. The CIFAR-100 dataset [26] consists of 60000 color images in 32 by 32 resolution; the number of classes is 100, resulting in 600 images per class. Similar to CIFAR-10, there are 50000 training images and 10000 test images. The implementation of the selected Transformer architectures was chosen to obtain an equalized comparative setting to highlight the effect of the chosen optimizers. For the ViT architecture, the chosen patch size was 4 with a token dimension of 256, and the number of layers chosen was 4 with 4 attention heads and an MLP dimension of 512. For the MLPMixer architecture, the chosen patch size was 4 with a token dimension of 256, and the number of layers chosen was 4 with a token-wise MLP dimension of 512 and a channel-wise MLP dimension of 512. For the ConvMixer architecture, the chosen patch size was 4, with a convolution kernel size of 8, and the number of channels for each layer was 64, and number of layers chosen was 4. All models were fitted with a training loop comprised of 100 epochs with a batch size of 128. Except for the LION optimizer experiments, which had a stable learning rate of 0.0001, all other experiments adopted the recommended learning rate of 0.001 [27]. The utilized software platform is Pytorch of version 1.13, supported by CUDA toolkit of version 11.6.2. The version used for Python programming language is 3.9.The available hardware system is a desktop personal computer with Intel i9-9900k CPU, 32 Gigabytes of system RAM, Nvidia RTX 2080ti GPU with 12 Gigabytes of RAM and running an UBUNTU 20 LTS operating system.

After performing the experimentation, it was recorded that for the Adagrad optimizer, the obtained test accuracy was very low on both chosen datasets, with the best test accuracy on CIFAR-10 being 52.98% obtained on the ConvMixer Transformer and the best test accuracy on CIFAR-100 being 26.16% obtained on the MLPMixer Transformer. For Adam optimizer, the obtained test accuracy of this optimizer was competitive on both chosen datasets, with the best test accuracy on CIFAR-10 being 79.67% obtained on ConvMixer Transformer and the best test accuracy on CIFAR-100 being 49.21% obtained on ConvMixer Transformer. For AdamW optimizer, the obtained test accuracy of this optimizer was more competitive on both chosen datasets, with the best test accuracy on CIFAR-10 being 80.42% obtained on ConvMixer Transformer and the best test accuracy on CIFAR-100 being 50.36% obtained on ConvMixer Transformer. Finally, for the LION optimizer, the obtained test accuracy of this optimizer was the most performing overall on both chosen datasets, with the best test accuracy on CIFAR-10 being 81.89% obtained on ConvMixer Transformer and the best test accuracy on CIFAR-100 being 53.1% obtained on ConvMixer Transformer. The obtained accuracy tables for each dataset are shown below (Tables 1 and 2).

Figure 1 shows the best accuracy of each optimizer obtained on the selected datasets, while Figs. 2 and 3 show, respectively, the accuracy and loss curves of the best overall architecture (ConvMixer) obtained on LION optimizer for both CIFAR-10 and CIFAR-100 datasets.

Based on the recorded metrics and graphs, the low performance of the Adagrad optimizer can be attributed to the lack of sufficient regularization provided in the update rule of this optimizer. Both Adam and AdamW enhance the performance of the optimization process due to the application of strong regularization terms in the update rule and more utilization of gradient information to obtain better convergence. The favorable performance of the LION optimizer that significantly improves on the previous designs can be attributed to using the sign of the regularized gradient as extra information in addition to the value of the previous weight in the update step. These extra terms distinguish this optimizer found by algorithmic search from comparable alternatives that were devised by human designers and indicate that

**Table 1** Experimentally obtained accuracy metric on CIFAR-10 dataset

| Optimizer | Learning rate | Test accuracy (ViT) (%) | Test accuracy (MLPMixer)(%) | Test accuracy (ConvMixer)(%) |
|---|---|---|---|---|
| Adagrad | 0.001 | 37.15 | 52.16 | 52.98 |
| Adam | 0.001 | 65.74 | 72.28 | 79.67 |
| AdamW | 0.001 | 64.08 | **74.1** | 80.42 |
| LION | 0.0001 | **66.13** | 72.82 | **81.89** |

**Table 2** Experimentally obtained accuracy metric on CIFAR-100 dataset

| Optimizer | Learning rate | Test accuracy (ViT) (%) | Test accuracy (MLPMixer)(%) | Test accuracy (ConvMixer)(%) |
|---|---|---|---|---|
| Adagrad | 0.001 | 12.71 | 26.16 | 20.12 |
| Adam | 0.001 | 31.21 | 38.33 | 49.21 |
| AdamW | 0.001 | 35.23 | **42.78** | 50.36 |
| LION | 0.0001 | **36.86** | 37.14 | **53.1** |

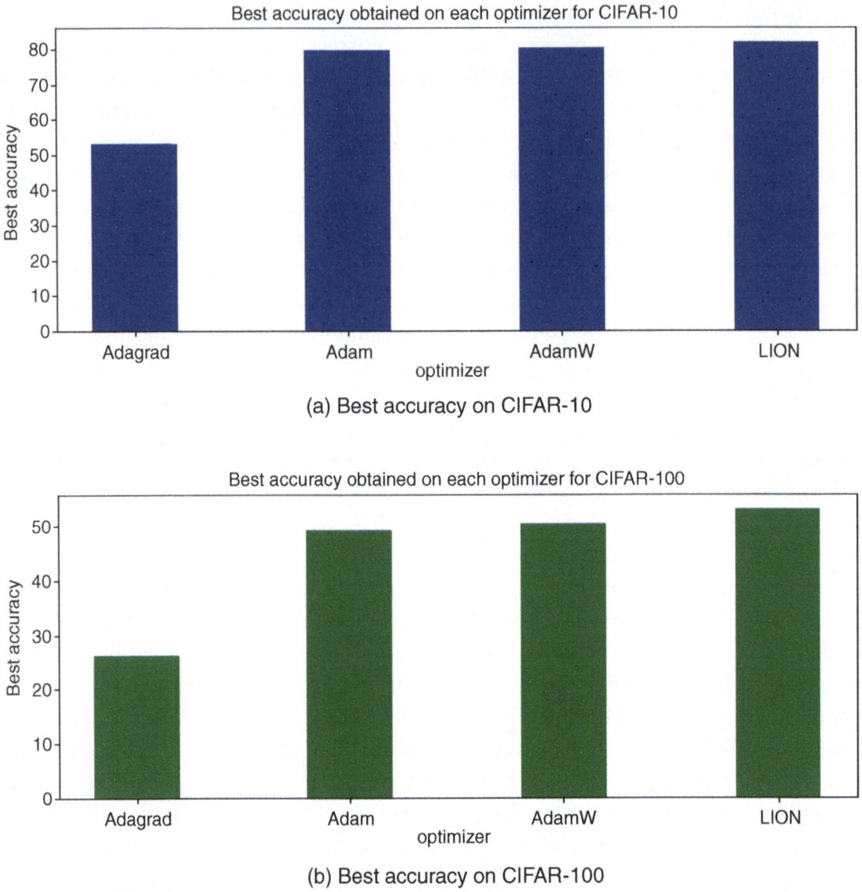


(a) Best accuracy on CIFAR-10

(b) Best accuracy on CIFAR-100

**Fig. 1** Best accuracy on selected datasets

more robust optimization schemes require more sophistication in the equations used in the weight update process.

## 5 Conclusion

Transformer architectures are new and highly generalized deep learning models proposed to be utilized in the computer vision domain. This necessitates identifying the proper optimizer that allows the highest degree of training ability. In this work, a comparative analysis of several optimizers has been performed and applied to three of the main Transformer architectures used in image processing, focusing on the task of image classification. It is identified that the novel and recently introduced

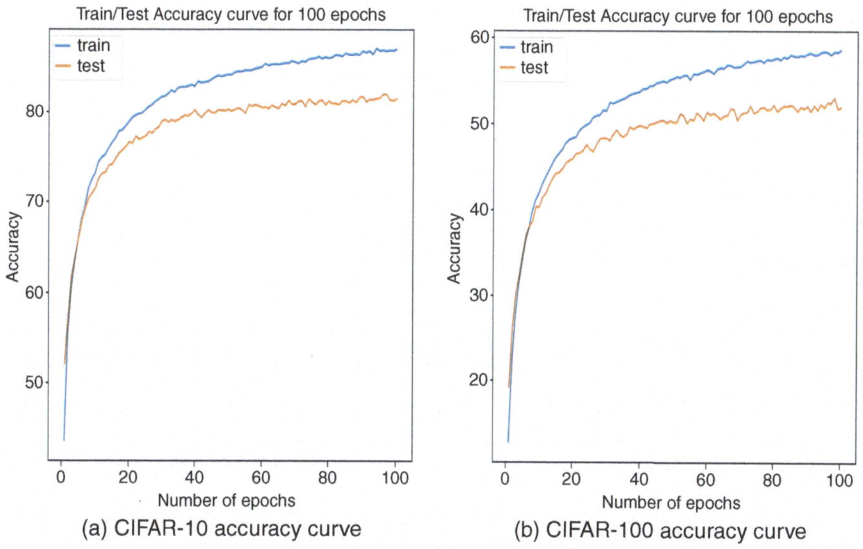


**Fig. 2** Accuracy curves for LION optimizer

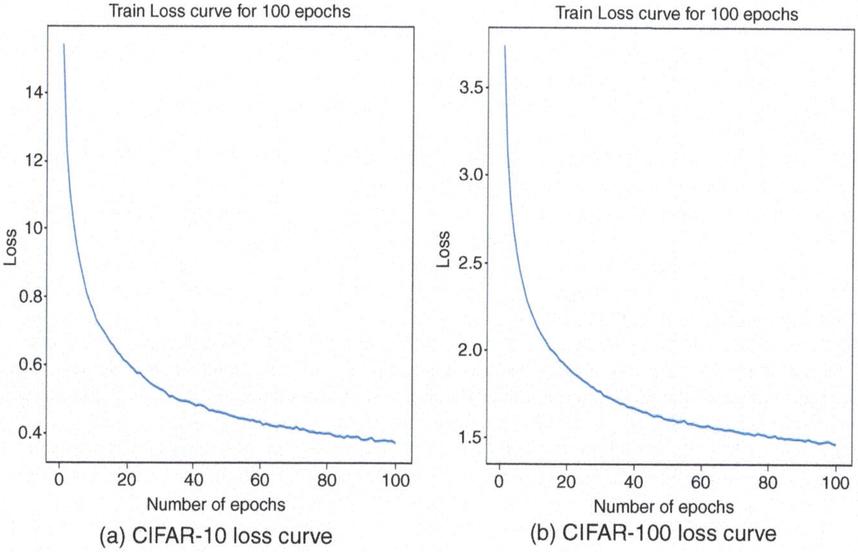


**Fig. 3** Loss curves for LION optimizer

optimizer LION is the most performing among the optimizer choices, and the study highly recommends it when utilizing Transformers in image processing tasks. It was also identified that Adagrad is the least performing optimizer across the board and recommends avoiding it when planning to adopt Transformers for computer vision experimentation. Future directions include investigating and uncovering the general qualities of optimizers that provide high performance in the fitting process of Transformers. Another venue of research is constructing the optimizer to require only forward passes and calculating proxy gradient estimations instead of explicit derivatives.

## References


1. Vaswani, A., Shazeer, N., Parmar, N., Uszkoreit, J., Jones, L., Gomez, A.N., Kaiser, Ł., Polosukhin, I.: Attention is all you need. In: Advances in Neural Information Processing Systems, vol. 30 (2017)
2. Dosovitskiy, A., Beyer, L., Kolesnikov, A., Weissenborn, D., Zhai, X., Unterthiner, T., Dehghani, M., Minderer, M., Heigold, G., Gelly, S., Uszkoreit, J., Houlsby, N.: An Image is Worth 16x16 Words: Transformers for Image Recognition at Scale. ArXiv abs/2010.11929 (2020)
3. Xu, P., Zhu, X., Clifton, D.A.: Multimodal learning with transformers: A survey. IEEE Trans. Pattern Anal. Mach. Intell. **45**, 12113–12132 (2023)
4. Bommasani, R., Hudson, D.A., Adeli, E., Altman, R., Arora, S., von Arx, S., Bernstein, M.S., et al.: On the opportunities and risks of foundation models. arXiv preprint arXiv:2108.07258 (2021)
5. Tolstikhin, I.O., Houlsby, N., Kolesnikov, A., Beyer, L., Zhai, X., Unterthiner, T., Yung, J., Keysers, D., Uszkoreit, J., Lucic, M., Dosovitskiy, A.: MLP-Mixer: An all-MLP Architecture for Vision. Neural Information Processing Systems (2021)
6. Trockman, A., Zico Kolter, J.: Patches Are All You Need? Trans. Mach. Learn. Res. **2023**, (2022)
7. Choi, D., Shallue, C.J., Nado, Z., Lee, J., Maddison, C.J., Dahl, G.E.: On empirical comparisons of optimizers for deep learning. arXiv preprint arXiv:1910.05446 (2019)
8. Wilson, A.C., Roelofs, R., Stern, M., Srebro, N., Recht, B.: The marginal value of adaptive gradient methods in machine learning. In: Advances in Neural Information Processing Systems, vol. 30, pp. 4148–4158. Curran Associates, Inc. (2017)
9. Schneider, F., Balles, L., Hennig, P.: DeepOBS: a deep learning optimizer benchmark suite. arXiv preprint arXiv:1903.05499 (2019)
10. Dogo, E.M., Afolabi, O.J., Nwulu, N.I., Twala, B., Aigbavboa, C.O.: A Comparative Analysis of Gradient Descent-Based Optimization Algorithms on Convolutional Neural Networks. In: Proceedings of the 2018 International Conference on Computational Techniques, Electronics and Mechanical Systems (CTEMS), Belgaum, India. 21–22 December 2018
11. Prilianti, K.R., Brotosudarmo, T.H.P., Anam, S., Suryanto, A.: Performance Comparison of the Convolutional Neural Network Optimizer for Photosynthetic Pigments Prediction on Plant Digital Image. AIP Publishing, University Park, MA, USA (2019)
12. Jangid, M., Srivastava, S.: Deep ConvNet with different stochastic optimizations for handwritten devanagari character. In: Proceedings of the IC4S 2017; Patong Phuket, Thailand, 11–12 October 2017, pp. 51–60
13. Swastika, W., Ariyanto, M.F., Setiawan, H., Irawan, P.L.T.: Appropriate CNN architecture and optimizer for vehicle type classification system on the toll road. J. Phys. Conf. Ser. **2019**, 1196:012044. doi:10.1088/1742-6596/1196/1/012044
14. Deng, P., Xu, K., Huang, H.: When CNNs Meet Vision Transformer: A Joint Framework for Remote Sensing Scene Classification. In: IEEE Geoscience and Remote Sensing Letters, vol. 19, pp. 1–5, 2022, Art no. 8020305

15. Yu, S., Ma, K., Bi, Q., Bian, C., Ning, M., He, N., Li, Y., Liu, H., Zheng, Y.: Mil-vt: Multiple instance learning enhanced vision transformer for fundus image classification. In: Proc. of the International Conference on Medical Image Computing and Computer-Assisted Intervention, pp. 45–54. Springer, Cham (2021)
16. Xue, Z., Tan, X., Yu, X., Liu, B., Yu, A., Zhang, P.: Deep hierarchical vision transformer for hyperspectral and LiDAR data classification. IEEE Trans. Image Process. **31**, 3095–3110 (2022)
17. Tanzi, L., Audisio, A., Cirrincione, G., Aprato, A., Vezzetti, E.: Vision Transformer for femur fracture classification. Injury (Apr. 2022)
18. Beal, J., Kim, E., Tzeng, E., Park, D.H., Zhai, A., Kislyuk, D.: Toward transformer-based object detection. arXiv preprint arXiv:2012.09958 (2020)
19. Li, Y., Xie, S., Chen, X., Dollar, P., He, K., Girshick, R.: Benchmarking detection transfer learning with vision transformers. arXiv preprint arXiv:2111.11429 (2021)
20. Hatamizadeh, A., Xu, Z., Yang, D., Li, W., Roth, H., Xu, D.: UNetFormer: A Unified Vision Transformer Model and Pre-Training Framework for 3D Medical Image Segmentation. arXiv preprint arXiv:2204.00631 (2022)
21. Yuan, H., Cai, Z., Zhou, H., Wang, Y., Chen, X.: TransAnomaly: video anomaly detection using video vision transformer. IEEE Access **9**, 123977–123986 (Aug. 2021)
22. Duchi, J., Hazan, E., Singer, Y.: Adaptive subgradient methods for online learning and stochastic optimization. J. Mach. Learn. Res. **12**(7), 2121–2159 (2011)
23. Kingma, D.P., Ba, J.: Adam: A method for stochastic optimization. arXiv preprint arXiv:1412.6980 (2014)
24. Loshchilov, I., Hutter, F.: Decoupled weight decay regularization. arXiv preprint arXiv:1711.05101 (2017)
25. Chen, X., Liang, C., Huang, D., Real, E., Wang, K., Liu, Y., Pham, H., et al.: Symbolic discovery of optimization algorithms. arXiv preprint arXiv:2302.06675 (2023)
26. Krizhevsky, A., Nair, V., Hinton, G.: Cifar-10 and cifar-100 datasets, vol. 6(1), (2009). https://www.cs.toronto.edu/kriz/cifar.html
27. Steiner, A., Kolesnikov, A., Zhai, X., Wightman, R., Uszkoreit, J., Beyer, L.: How to train your vit? data, augmentation, and regularization in vision transformers (2021)

# Comparing LSTM and Transformer for Video Depth Estimation

**Rozhin Fani and Berke Gur**

## 1 Introduction

Accurate depth estimation is critical for 3D scene understanding in applications such as robotics, autonomous vehicles, and augmented reality. While progress is made in monocular depth prediction using convolutional neural networks (CNNs) [1], estimates from single images remain limited by factors including noise, occlusion, and lack of context.

Several works have sought to address these limitations by incorporating temporal information across video frames. Techniques such as extracting optical flow between frames [2] and aggregating context using recurrent LSTM networks have shown potential for improving depth consistency and accuracy [3]. However, recurrent neural networks (RNNs) exhibit challenges in capturing long-range dependencies across sequences [4]. This work aims to overcome these limitations by exploiting temporal information from consecutive frames.

Recently introduced Transformer networks offer a promising alternative for sequence modeling. By relying on self-attention, Transformers can directly model global context and long-range dependencies [5, 6]. Owing to such abilities, Transformers are utilized in fields like natural language processing [7], time series analysis, and computer vision [8].

This research is supported by The Scientific and Technological Research Council of Türkiye (TÜBİTAK) through the project entitled "Development of autonomous warehouse multi-robotic system (Ro-PATH)" (grant no. 220N341).

R. Fani (✉) · B. Gur
Bahcesehir University, İstanbul, Türkiye
e-mail: rozhin.fani@bahcesehir.edu.tr; berke.gur@bau.edu.tr

 
Ö. M. Gül et al. (eds.), *7th EAI International Conference on Robotic Sensor Networks*, EAI/Springer Innovations in Communication and Computing,
https://doi.org/10.1007/978-3-031-64495-5_7

In this work, a Transformer-based pipeline for video depth estimation, integrating optical flow extraction with self-attention-based temporal aggregation, is presented. Experiments demonstrate quantitatively that the proposed approach outperforms recurrent LSTM networks for indoor scene depth prediction. The results provide valuable insights into the advantages of Transformer encoders for spatiotemporal modeling compared to prior RNN techniques. The key contributions of the work are twofold: (1) developing an end-to-end depth pipeline with optical flow extraction and Transformer-based sequence encoding and (2) conducting a comparative study of Transformer vs LSTM network for temporal context aggregation.

The remainder of the paper is structured as follows. A concise review of monocular depth estimation, optical flow, and sequence modeling is presented in Sect. 2. The proposed modeling and depth estimation methods are explained in detail in Sect. 3. In Sects. 4 and 5, specifics of the model training, experiments, and key results are presented. The paper is concluded in Sect. 6 with conclusions and future work.

## 2 Related Work

A concise review of relevant research is presented in this section. Existing literature is grouped into four categories including monocular depth estimation, optical flow, sequence modeling, and transformers.

### *2.1 Monocular Depth Estimation*

Estimating scene depth from single images is a long-standing problem in computer vision with many applications in robotics, autonomous vehicles, and augmented reality. Earlier works relied on handcrafted features and graphical model inference to predict depth maps from monocular images [9]. With the advent of deep learning, CNN-based methods have yielded significant improvement in monocular depth estimation. One of the pioneering works by Eigen et al. [1] demonstrated the potential of CNNs by developing multi-scale architectures. This laid the foundation for subsequent improvements through techniques such as fully convolutional residual networks [10], which increased depth accuracy through skip connections. Additional improvements have emerged from formulating depth estimation as a regression problem and using coarse-to-fine predictions [11]. More recent works have enhanced monocular depth through semi-supervised learning [12]. These methods provide unlabeled cues through reconstruction losses. While these approaches represent significant progress, robust monocular depth estimates from single images still remain illusive due to noise and lack of global context.

## 2.2 Optical Flow

Optical flow estimation is a fundamental technique in computer vision that plays a vital role in depth map prediction. By analyzing the apparent motion of pixels between consecutive frames, optical flow provides valuable cues for inferring depth information. This has led to the popularity of optical flow-based methods for depth map prediction, as they leverage the temporal information present in sequential images to improve depth estimation accuracy.

One prominent optical flow algorithm used in depth map prediction is the Farneback method [2]. This method is widely adopted due to its efficiency and robustness. It provides low computational cost and fast processing, making it suitable for real-time robotics applications. Additionally, it maintains accuracy in the presence of image disturbances, enabling accurate optical flow estimation.

The Farneback method utilizes a polynomial expansion to model local motion and employs a pyramidal approach to effectively handle large displacements [2]. By considering intensity changes between frames, the algorithm estimates dense motion fields, which can then be utilized for depth map prediction. While the Farneback method assumes brightness constancy and has certain limitations in handling large displacements, its real-time performance and noise robustness have contributed to its widespread use in depth map prediction tasks.

## 2.3 Sequence Modeling

Depth map prediction using image sequences is an active research area in computer vision. Several studies have explored the use of RNNs and LSTM models to leverage temporal information for improved depth estimation. For example, Behl et al. [13] demonstrated the effectiveness of extracting optical flow between frames and using LSTM networks to aggregate context for enhanced depth estimates.

However, RNNs and LSTMs have certain drawbacks in the context of depth map prediction. One significant limitation is the difficulty these methods experience in capturing long-range dependencies. Backpropagation through time, which is commonly used for training RNNs and LSTMs, can suffer from vanishing or exploding gradients [6]. As a result, the models may struggle to capture complex temporal patterns and produce less reliable depth estimates.

To overcome these limitations, researchers have explored alternative models such as Transformer networks, which utilize self-attention mechanisms to directly model global dependencies in sequences [14]. Transformers have shown promise in various sequence-based tasks and have the potential to better exploit long-range temporal cues for depth estimation.

### 2.4 *Transformers*

Transformers are a type of advanced neural network architecture that has gained significant attention in recent years. Originally developed for natural language processing tasks [14], Transformers have proven to be highly effective in various computer vision tasks, including monocular depth estimation from a single image.

Yang et al. [15] presented a Transformer architecture for continuous pixel prediction. The multilevel attention modeled spatial dependencies and improved structured depth estimates over CNNs. A limitation is that it only leverages single image context. Another limitation is that the proposed architecture can be sensitive to noise in the input image. This is because transformers are trained to learn long-range dependencies in the image, and noise can disrupt these dependencies. Using image sequences and their optical flow could provide temporal cues to further enhance accuracy and consistency.

Chang et al. [16] proposed a monocular depth Transformer with attention supervision. It significantly improved accuracy by focusing on geometric regions. However, the proposed method lacks temporal modeling capabilities. Incorporating sequences could help resolve ambiguities and occlusions over time. Jeeveswaran et al. [17] introduced Vision Transformers for dense tasks like depth estimation. Concatenating self-attention features boosted performance by enabling better propagation of spatial signals. However, it does not leverage temporal information. Modeling sequences could allow aggregating useful cues across frames to further improve accuracy.

While current Transformer networks have demonstrated strong performance for monocular depth tasks by encoding spatial context, they lack temporal modeling capabilities. The work described in this paper aims to address these limitations by exploring Transformer architectures that aggregate information across image sequences for enhanced video depth estimation.

## 3 Proposed Method

This work compares two architectures for monocular depth estimation from image sequences. Both models use an Encoder network to extract compact features representing optical flow between frames. These features capture low-level motion cues. To aggregate temporal context, Model A passes the Encoder outputs to an LSTM, while Model B uses a Transformer. The LSTM or Transformer provides a sequence-level representation that summarizes the contextual information across frames. This context vector is decoded by a encoder-decoder style network with skipping connections to predict the depth map for the central frame. Both models are trained end to end with losses on the predicted and ground truth depth. Extensive evaluation is done by computing depth prediction metrics on a held-out validation set after each training epoch. This allows quantitative comparison

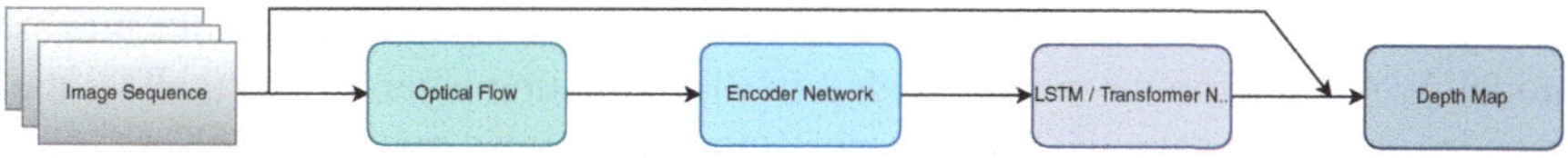


**Fig. 1** The schematic of the depth net pipeline

between the LSTM and Transformer architectures for temporal feature aggregation in monocular depth estimation. Both methods are evaluated to determine which approach better leverages sequences to improve single-frame depth prediction over individual images. Figure 1 shows an overview of our depth estimation pipeline that uses either the LSTM or the Transformer Network.

## *3.1 Optical Flow*

Farneback algorithm is used to estimate dense optical flow between grayscale frames. This approximates the pixel motion between frames using a polynomial expansion. Flow vectors are stacked to create two-channel optical flow images. The local polynomial approximation, defined as

$$f(\mathbf{x}) \approx \mathbf{x}^T \mathbf{A}\mathbf{x} + \mathbf{b}^T \mathbf{x} + \mathbf{c} \tag{1}$$

fits a quadratic polynomial $f(\mathbf{x})$ to each pixel neighborhood. A displacement constraint

$$\mathbf{Ad} = \Delta\mathbf{b} \tag{2}$$

relates the polynomial coefficients $\mathbf{A}$ and $\mathbf{b}$ before and after displacement $\mathbf{d}$ to estimate $\mathbf{d}$. Neighborhood flow estimation

$$\mathbf{d} = (\mathbf{WA}^T\mathbf{A})^{-1}\mathbf{WA}^T \Delta\mathbf{b} \tag{3}$$

estimates the displacement by minimizing errors across a neighborhood with weights $\mathbf{W}$.

## *3.2 Temporal Feature Extraction*

The first model (Model A) that incorporates an LSTM for extracting the spatiotemporal features. In this model, an Encoder model with convolutional layers compresses the optical flow images into $1 \times 128$ sized feature vectors. A single-layer LSTM with 512 hidden units aggregates the Encoder features across the sequence. The LSTM output is decoded by the DepthNet module to predict depth. The second model (Model B) also includes an Encoder model that has the same architecture as Model A. A Transformer module with 8-headed attention layers contextualizes the Encoder outputs. Again, the Transformer output is decoded by the DepthNet.

### *3.3 DepthNet*

The DepthNet module is designed to take the context vector from the LSTM or Transformer and decode it into a full-resolution depth map prediction. It has a U-Net style encoder-decoder architecture with convolutional layers. The encoder part of DepthNet is a convolutional neural network that encodes the input image into a set of feature maps. It consists of several convolutional layers, each with a $5 \times 5$ kernel and stride of 2, which downsamples the feature maps to capture higher-level representations while reducing spatial dimensions. Batch normalization and ReLU activation are applied after each convolution.

The decoder part concatenates the context vector with encoder features and applies $3 \times 3$ convolutions followed by batch normalization and ReLU activation to fuse the global and local information. Transposed convolutions are used to upsample the feature maps. Skip connections from the encoder are incorporated to retain spatial details. This allows the network to use both the compact global context and fine-grained local features from the Encoder to reconstruct the detailed depth structure accurately. Finally, a last transposed convolution layer with a $2 \times 2$ kernel reconstructs the full-resolution depth map prediction from the fused encoder-decoder representations.

### *3.4 Loss Function and Training*

Models are trained with $L^1$ loss on predicted and ground truth depth, plus a smoothness regularization loss. The Adam optimizer is used with learning rate decayed exponentially every 15.000 iterations.

### *3.5 Evaluation*

Quantitative depth prediction metrics such as absolute relative error (AbsRel), squared relative error (SqRel), root mean squared error (RMSE), and accuracy with threshold $\delta_\epsilon$ are computed on the validation set. These metrics are, in order

$$\text{AbsRel} = \frac{1}{|T|} \sum_{i \in T} \left( \frac{|\hat{y}_i - y_i| + 10^{-4}}{y_i + 10^{-4}} \right) \tag{4}$$

$$\text{SqRel} = \frac{1}{|T|} \sum_{i \in T} \frac{(\hat{y}_i - y_i)^2 + 10^{-4}}{y_i + 10^{-4}} \tag{5}$$

$$\text{RMSE} = \sqrt{\frac{1}{|T|} \sum_{i \in T} (\hat{y}_i - y_i)^2} \tag{6}$$

where $\hat{y}_i$ and $y_i$ represent the predicted and ground truth depth, respectively, for pixel $i$. In addition, $T$ represents the total number of pixels.

The threshold accuracy formula checks each pixel $i$ and computes the ratio between predicted and ground truth depth according to

$$\tau_i = \max\left(\frac{\hat{y}_i}{y_i}, \frac{y_i}{\hat{y}_i}\right) \tag{7}$$

To compute the threshold accuracy, the computed ratio $\tau_i$ is compared it to the threshold ratio $\epsilon$

$$\delta_\epsilon = \frac{1}{|T|} \sum_{i \in T} \delta(\tau_i < \epsilon) \tag{8}$$

This allows assessing depth prediction accuracy across different precision levels.

# 4 Validation Results

## *4.1 Data*

The dataset consists of 23.158 RGB images and corresponding depth maps from KITTI Dataset with Eigen split [1]. The images were augmented with random horizontal flips, vertical flips, and rotations. The augmented images and depth maps are converted to Torch Tensors and normalized to [0,1] range.

## *4.2 Training*

The end-to-end networks with Transformer and LSTM blocks were trained on an NVIDIA RTX 3060 GPU. Their training times were not significantly different. Two models (i.e., LSTM and Transformer architectures) are evaluated for monocular depth estimation. Both leverage temporal context from sequences to improve per-frame depth prediction.

The LSTM model achieves an absolute relative error (AbsRel) of 0.079 and RMSE of 0.140 on the validation set. The $\delta < 1.25$ threshold is 0.900, with

**Table 1** The error metrics obtained for LSTM and transformer models

| Model | AbsRel | SqRel | RMSE | $\delta < 1.25$ | $\delta < 1.25^2$ | $\delta < 1.25^3$ |
|---|---|---|---|---|---|---|
| LSTM | 0.079 | 0.020 | 0.140 | 0.900 | 0.956 | 0.981 |
| Transformer | 0.012 | 0.001 | 0.027 | 0.998 | 1.000 | 1.000 |

$\delta < 1.25^2$ at 0.956 and $\delta < 1.25^3$ being 0.981. These metrics indicate the LSTM produces accurate depth estimates, with most predictions within a 1.25 ratio of ground truth. The low absolute relative error shows depths are close to true values in absolute terms.

The Transformer model obtains significantly better metrics than the LSTM. The AbsRel error and the RMSE are 0.012 and 0.027, respectively, after training. The $\delta < 1.25$ score is 0.998, with $\delta < 1.25^2$ and $\delta < 1.25^3$ both reaching 1.000. This demonstrates the Transformer can estimate depth with very high accuracy. The extremely low errors and near perfect thresholds validate its effectiveness at leveraging temporal context.

The Transformer significantly outperforms the LSTM on all metrics, decreasing the absolute relative error by six times and the RMSE by five times. The results demonstrate the Transformer architecture is superior to LSTM for aggregating temporal context from sequences for monocular depth estimation. The self-attention mechanism better models long-range dependencies across frames compared to the LSTM's limited memory. This allows the Transformer to effectively leverage multi-frame data to produce highly accurate per-frame depth maps (Table 1).

## 5 Qualitative Results

An example of typical results obtained for depth estimation is presented in this section. Representative three consecutive sample images from the validation dataset are shown in Fig. 2. The associated ground truth depth maps of each image is also provided in Fig. 3. These serve as a reference for gauging how well the models reconstructed the true depth structure.

The predicted depth maps from the LSTM and Transformer models are shown in Figs. 4 and 5, respectively. When comparing these to the Transformer results, the borders and object boundaries seem blurrier or less defined. Visually comparing the Transformer predictions of Fig. 5 to the ground truths presented in Fig. 3, it can be concluded that Transformer estimates appear to more accurately capture the finer details like discrete steps and crisp boundaries present in the scene.

In contrast, the LSTM predictions (see Fig. 4) seem to lack some of these fine-grained structural elements when referenced against the ground truths. However, in Fig. 4c, the network seems to fail completely at estimating depth for this scene. Rather than capturing the discrete steps and variations in the ground truth Fig. 3c, the LSTM prediction contains no meaningful depth information. This suggests the

(a) (b) (c)

**Fig. 2** Typical consecutive RGB images from the KITTI dataset

(a) (b) (c)

**Fig. 3** Ground truth depth maps

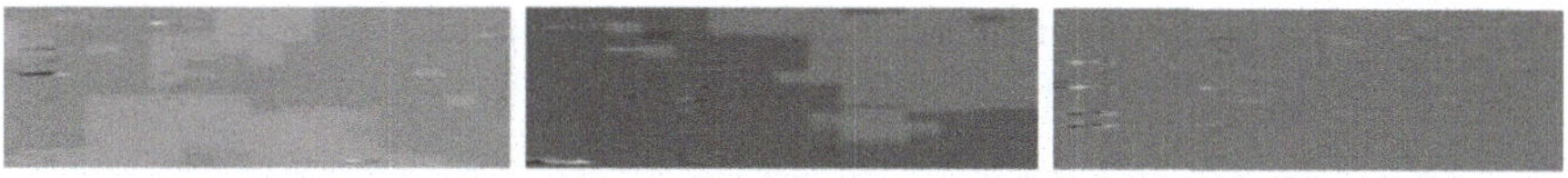

(a) (b) (c)

**Fig. 4** Depth maps obtained with the LSTM model

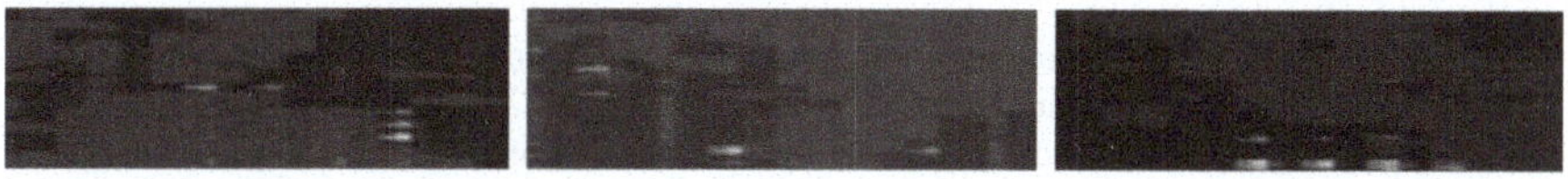

(a) (b) (c)

**Fig. 5** Depth maps obtained with the transformer model

LSTM struggles more than the Transformer with certain complex input cases. Its limited receptive field may prevent it from aggregating enough contextual cues across frames to infer the 3D scene structure.

## 6 Conclusion

This work presented a comparative study between LSTM and Transformer architectures for temporal context aggregation in monocular video depth estimation. The core analysis examined which model better leverages multi-frame cues to enhance per-frame depth prediction over individual images.

The results demonstrate the Transformer provides superior sequence modeling capabilities compared to recurrent LSTM networks. On the KITTI dataset, the Transformer reduced the depth error metrics by over six times compared to the LSTM model. The attention mechanism enables the Transformer to propagate signals directly across long frame sequences. In contrast, LSTM struggles to capture such long-range dependencies due to issues like vanishing gradients.

The findings highlight the advantages of self-attention for temporal feature aggregation. By allowing interactions between all sequence elements, the Transformer can integrate information from wider context. This helps resolve ambiguities and improves depth consistency across frames.

This work helps provide guidance on effective sequence architectures for spatiotemporal perception problems in robotics and related fields. It validates the Transformer as a promising model for video-based depth prediction. Directions for future work include exploring different attention formulations and combining recurrent elements with self-attention. The findings provide intriguing possibilities for applying Transformer networks to other robotic perception tasks such as visual odometry, 3D pose estimation, and SLAM that involve temporal integration of signals. The Transformer's sequence modeling strengths could allow aggregating cues across frames to achieve more robust ego-motion estimation and mapping. This work helps provide a foundation for leveraging self-attention models to enhance spatiotemporal understanding in robotics systems.

## References

1. Eigen, D., Puhrsch, C., Fergus, R.: Depth map prediction from a single image using a multi-scale deep network. Adv. Neural Inf. Process. Syst. **27**, 2366–2374 (2014)
2. Farnebäck, G.: Two-frame motion estimation based on polynomial expansion. In: Proceedings of the 13th Scandinavian Conference on Image Analysis (SCIA), pp. 363–370, Halmstad (2003)
3. Hochreiter, S., Bengio, Y., Farosconi, P.: Gradient flow in recurrent nets: the difficulty of learning long-term dependencies. In: A Field Guide to Dynamical Recurrent Neural Networks. IEEE Press, Piscataway (2001)
4. Graves, A.: Sequence transduction with recurrent neural networks. ArXiv: 1211.3711 (2012)
5. Hochreiter, S., Schmidhuber, J.: Long short-term memory. Neural Comput. **9**(8), 1735–1780 (1997)
6. Bengio, Y., Simard, P., Farosconi, P.: Learning long-term dependencies with gradient descent is difficult. IEEE Trans. Neural Netw. **5**(2), 157–166 (1994)
7. Devlin, J., Chang, M.-W., Lee, K., Toutanova, K.: BERT: Pre-training of deep bidirectional transformers for language understanding. In: Proceedings of the Conference of the North American Chapter of the Association for Computational Linguistics: Human Language Technologies, vol. 1, pp. 4171–4186, Minneapolis, MN (2019)
8. Carion, N., Massa, F., Synnaeve, G., Usunier, N., Kirillov, A., Zagoruyko, S.: End-to-end object detection with transformers. In: Proceedings of the European Conference on Computer Vision (ECCV). Spinger Lecture Notes in Computer Science, vol. 12346. pp. 213–229 (2020)
9. Saxena, A., Jain, A., Sener, O., Jami, A., Misra, D.K., Koppula, H.S.: RoboBrain: Large-scale knowledge engine for robots. ArXiv: 1412.0691 (2014)
10. Laina, I., Rupprecht, C., Belagiannis, V., Tombari, F., Navab, N.: Deeper depth prediction with fully convolutional residual networks. In: Fourth International Conference on 3D Vision, pp. 239–248, Stanford, CA (2016)
11. Fu, H., Gong, M., Gong, M., Wang, C., Batmanghelich, K., Tao, D.: Deep ordinal regression network for monocular depth estimation. In: Proceedings of the IEEE Conference on Computer Vision and Pattern Recognition (CVPR), pp. 2002–2011 (2018)

12. Kuznietsov, Y., Stückler, J., Leibe, B.: Semi-supervised deep learning for monocular depth map prediction. In: IEEE Conference on Computer Vision and Pattern Recognition (CVPR), pp. 2215–2223 (2017)
13. Behl, A., Jafari, O.H., Mustikovela, S.K., Alhaija, H.A., Rother, C., Geiger, A.: Bounding boxes, segmentations and object coordinates: How important is recognition for 3D scene flow estimation in autonomous driving scenarios? In: IEEE International Conference on Computer Vision (ICCV), pp. 2593–2602, Venice (2017)
14. Vaswani, A., Jones, L., Shazeer, N., Parmar, N., Gomez, A. N., Uszkoreit, J., Kaiser, Ł., Polosukhin, I.: Attention is all you need. Adv. Neural Inf. Process. Syst. **30**, 5998–6008 (2017)
15. Yang, G., Tang, H., Ding, M., Sebe, N., Ricci, E.: Transformer-based attention networks for continuous pixel-wise prediction. In: Proceedings of the IEEE/CVF International Conference on Computer Vision (ICCV), pp. 16249–16259 (2021)
16. Chang, W., Zhang, Y., Xiong, Z: Transformer-based monocular depth estimation with attention supervision. In: Proceedings of the 32nd British Machine Vision Conference (BMVC), vol. 6, p. 7 (2021)
17. Jeeveswaran, K., Kathiresan, S., Varma, A., Magdy, O., Zonooz, B., Arani, E.: A comprehensive study of vision transformers on dense prediction tasks. In: Proceedings of the 17th International Joint Conference on Computer Vision, Imaging and Computer Graphics Theory and Applications (VISAPP), vol. 4, pp. 213–223 (2022)

# Style Transfer to Enhance Data Augmentation for Facial Action Unit Detection

**Duygu Cakir and Nafiz Arica**

## 1 Introduction

Data augmentation techniques, which involve generating new training samples by applying various transformations or perturbations to existing data, are commonly used in computer vision tasks such as face-related tasks [1], medical image analysis [2], autonomous driving [3, 4], underwater robotics [5], construction safety [6], and even agricultural robotics for crop monitoring [7]. These augmentation techniques can be applied to improve the performance and robustness of machine learning models used in object detection, object recognition, and scene understanding.

By augmenting the available training data, models can be exposed to a wider range of variations and scenarios, enabling them to generalize better to real-world conditions. For instance, in robotic vision tasks, data augmentation techniques can be used to simulate different lighting conditions, camera perspectives, occlusions, and object deformations, allowing the model to learn to handle these variations and improve its performance in diverse environments.

Furthermore, data augmentation addresses the challenge of limited training data. Collecting large-scale and diverse datasets for robotics applications can be time-consuming, expensive, or even impractical in certain scenarios. Data augmentation provides a cost-effective solution by artificially expanding the training dataset, reducing the risk of overfitting and improving the model's ability to generalize. The use of data augmentation techniques in image recognition tasks has been

D. Cakir (✉)
Bahcesehir University, İstanbul, Türkiye
e-mail: duygu.cakir@bau.edu.tr

N. Arica
Piri Reis University, İstanbul, Türkiye
e-mail: nafiz.arica@pirireis.edu.tr



Ö. M. Gül et al. (eds.), *7th EAI International Conference on Robotic Sensor Networks*, EAI/Springer Innovations in Communication and Computing,
https://doi.org/10.1007/978-3-031-64495-5_8

widely recognized as crucial for improving model performance and addressing the limitations of limited training data [8].

The goal of face analysis is to extract as much information as possible to facilitate in applications such as automation [9], medicine [10–12], security [13, 14], robotics [15], driver monitoring [16], and even mobile phone apps [17] such as aging, putting on makeup, changing the style, measuring the attractiveness, etc. Facial Action Units (AU) are the micro-mimic movements of the face that are used to analyze and extract the facial information of people [18] to be used in a wide range of applications.

In the context of AU recognition, data augmentation techniques are commonly used to increase the diversity and size of the training dataset, leading to improved model performance. However, traditional data augmentation methods such as rotation, scaling, and translation may not effectively capture the complex variations in facial movements. Therefore, there is a need for innovative approaches that can enhance data augmentation specifically for facial expression datasets [19–21].

This paper proposes an approach that uses style transfer for enhancing data augmentation in facial image recognition tasks. The experiments have been carried on using two spontaneous datasets (DISFA [22], BP4D [23]) and one in-the-wild dataset (EmotioNet [24]) by transferring two different styles over the images. The proposed method aims to create additional training samples that capture the full range of facial movements, including subtle variations and occluded regions.

The main contributions of this paper are as follows: (1) integrating style transfer into the data augmentation process to synthesize facial images, (2) leveraging style transfer to improve the accuracy and robustness of facial action unit detection models, and (3) overcoming the challenge of limited and uneven datasets in face-related tasks.

The rest of the paper is organized as follows: Sect. 2 presents the proposed method in detail on how style transfer has been employed for data augmentation. Section 3 presents the results and analysis. Finally, Sect. 4 concludes the paper and discusses future research directions.

## 2 Methodology

The methodological part of this study examines the main factors underlying the analysis of the proposed method used to provide data enhancement for critical tasks related to facial action unit (AU) detection. It is widely accepted that accurate AU detection plays an important role in various industries. To pursue the development of effective and robust AU detection models under investigation, a new approach has been proposed that combines the powerful capabilities of style transfer with the aim of augmenting the faces on the training dataset.

Providing the details of the selected datasets, implementation details, and experimental settings, this section aims to clarify the methodology used to derive the proposed method and sets the stage for an in-depth analysis in the results and insights of the next section.

## 2.1 Database Setup

In the experiments of the study, the identical datasets employed by Cakir and Arica (2023) [25] have been utilized: (1) DISFA [22], BP4D [23], and (3) EmotioNet [24]. Each of these datasets are manually labelled by experts and contain frame level AU labelling on 2D RGB frames. By maintaining the consistent use of these datasets, it has been ensured that the data foundation remains unchanged. This not only allows for a direct comparison with the results presented in the earlier research but also facilitates an extension of the existing framework. Figure 1 contains samples of three different AUs belonging to these three datasets.

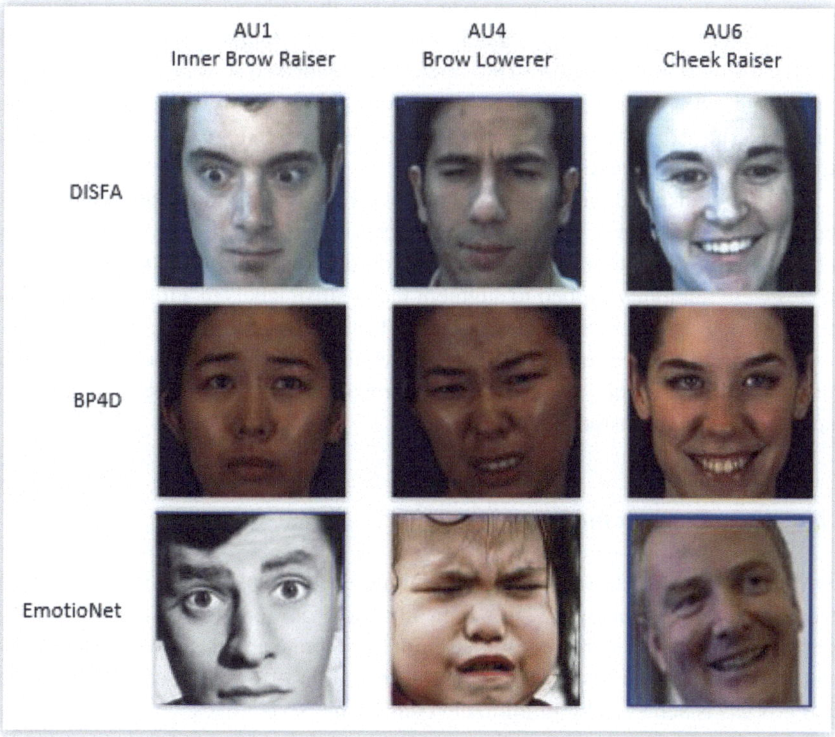


**Fig. 1** Three different Action Unit (AU) categories' positively labeled samples belonging to the three datasets that are examined in this study

As can be seen from Fig. 1, DISFA contains faces recorded from the front mostly without any occlusions or rotations, whereas BP4D has occlusion, lighting issues, and pitch-yaw angle alterations, making it a bit more challenging than DISFA; however, they are both lab-controlled datasets. On the other hand, EmotioNet is a dataset collected from the Internet and not lab controlled. It has many variations of lighting, ages, ethnicity, coloring, scene, background, etc. It is by far the most compelling public dataset in facial analysis as far as we are concerned.

## 2.2 *Implementation Details*

For each facial image in each dataset, faces are cropped using the Viola Jones algorithm [26] which are then resized to 224 × 224 preserving all three channels in all datasets.

Before and after employing the style transfer algorithm, consistent implementation took place within a controlled environment to ensure a fair comparison of the results. Prior to the application of style transfer, the only augmentation applied to the facial images involved a horizontal flip, effectively doubling the number of both positive and negative samples. This operation resulted in a total of 1600 positive images and 3200 negative samples, which were subsequently subjected to various transfer learning (TL) techniques prior to the style transfer augmentation to compare the accuracies before and after.

The style transfer algorithm was then applied exclusively to the positive images, generating an additional 500 synthetic positive images and thereby expanding the dataset to a total of 1300 positive samples. Additionally, 300 of the original positive images were randomly flipped to augment the positive labels to a total of 1600. The negative class remained at 3200 samples, collected using the same parameters.

The subsequent sections provide comprehensive details on the specific style transfer algorithm employed, and the transfer learning methods applied both before and after the style transfer augmentation.

### 2.2.1 Style Transfer for Image Synthesis

Style transfer is a technique used to combine the content of one image with the artistic style of another image. It is typically used for creating artistic and visually appealing images. Style transfer models aim to generate an image that preserves the content of a content image while adopting the style of a reference (style) image. This process is achieved by minimizing content and style losses.

For style transfer, ResNet, a deep convolutional neural network known for its ability to extract highly discriminative and abstract feature representations from images, is employed. The ResNet architecture incorporates residual blocks, enabling the training of deep networks more effectively.

**Fig. 2** (**a**) Vincent van Gogh, "Starry Night," 1889, Museum of Modern Art, New York, USA and (**b**) Georges Seurat, "View of Fort Samson," 1885, Hermitage Museum, Saint Petersburg, Russia

The foundation of the neural style transfer technique relies on two primary loss functions:

The total loss of the style transfer model is a linear combination of the content loss (the mean squared error (MSE) between feature maps extracted from the last convolutional layer of the ResNet) and the style loss (calculated by comparing the Gram matrices of feature maps), each scaled by a respective weighting factor.

To create the synthesized image, Adam optimization is applied to minimize the total loss function. The iterative optimization process fine-tunes the pixel values of the generated image to minimize the overall loss (Fig. 2).

This augmentation process introduced diversity into the dataset, enabling the transfer learning algorithms to generalize better and improve their performance in facial action unit detection.

### 2.2.2 Transfer Learning Base Models

The proposed method is an approach to prove a specific type of data augmentation and its effects on the model accuracy, and it has been seen more advantageous to employ many training models as the base model instead of a single method in order to fairly say that the proposed method successfully increases the expression recognition accuracies.

Since the proposed method had similar constraints, transfer learning (TL) using imagenet [27] weights with the same models and experimental settings have been employed as by the experiments applied on face masks [28].

While implementing the proposed algorithm, threefold cross validation has been applied and the average is taken for the AU accuracy.

## 3 Results

Two unique style images were employed for style transfer across these datasets. The first, "Starry Night," is characterized by its vibrant and intricate patterns, making it a highly complex and colorful source of stylistic influence. The second, "View of Fort Samson," is comparatively less intricate, providing a simpler stylistic influence. Tables 1, 2, and 3 display the applied TL algorithms before augmenting faces with style transfer and after the augmentation by transferring 2 different image styles to the dataset: (1) *The Starry Night (by Vincent van Gogh)* and (2) *View of Fort Samson (by Georges Seurat)*. Figure 3 contains samples of the original images as well as the artistic styles that are transferred to the images. Results on three datasets indicate that using style transfer as an augmentation method increases the AU detection scores under certain conditions:

**Table 1** Comparison of the popular TL methods' and the proposed techniques' $F_1$ scores and their averages on DISFA, trained with the original images and trained after both "The Starry Night" and "The View of Fort Samson" are transferred to the dataset as an augmentation method. Bolds indicate the increasing scores among each TL

| Method | 1 | 2 | 4 | 6 | 9 | 12 | 25 | 26 | Avg. |
|---|---|---|---|---|---|---|---|---|---|
| Xception | 88.9 | 93.4 | 90.0 | 93.1 | 95.6 | 95.1 | 91.1 | 88.8 | **92.0** |
| Xception + Starry Night | 88.3 | 91.9 | 84.6 | 93.6 | 90.6 | 93.5 | 92.5 | 88.1 | 90.4 |
| Xception + Fort Samson | 89.8 | 91.8 | 91.3 | 93.6 | 89.3 | 90.9 | 95.3 | 86.3 | 91.0 |
| InceptionV3 | 90.4 | 95.2 | 92.0 | 93.0 | 93.6 | 93.5 | 89.6 | 87.8 | **91.9** |
| InceptionV3 + Starry Night | 90.3 | 91.1 | 88.1 | 93.0 | 93.1 | 93.2 | 88.4 | 85.4 | 90.3 |
| InceptionV3 + Fort Samson | 92.4 | 92.1 | 90.9 | 92.0 | 91.0 | 91.5 | 88.0 | 83.3 | 90.2 |
| VGG16 | 86.9 | 93.4 | 91.1 | 92.8 | 89.7 | 92.7 | 83.5 | 90.9 | **90.1** |
| VGG16 + Starry Night | 88.1 | 89.0 | 85.8 | 90.0 | 90.2 | 90.0 | 86.7 | 81.2 | 87.6 |
| VGG16 + Fort Samson | 88.8 | 90.9 | 88.5 | 91.1 | 88.4 | 91.0 | 89.9 | 83.8 | 89.1 |
| VGG19 | 85.8 | 92.1 | 86.9 | 89.5 | 86.2 | 88.9 | 82.9 | 88.1 | **87.6** |
| VGG19 + Starry Night | 84.6 | 88.2 | 77.8 | 86.1 | 87.4 | 83.8 | 79.6 | 78.8 | 83.3 |
| VGG19 + Fort Samson | 85.1 | 88.7 | 84.7 | 87.7 | 85.4 | 85.0 | 86.8 | 84.5 | 86.0 |
| MobileNet | 90.0 | 94.1 | 91.0 | 94.4 | 92.5 | 95.9 | 91.2 | 90.6 | **92.5** |
| MobileNet + Starry Night | 86.5 | 91.1 | 88.8 | 95.4 | 92.2 | 94.7 | 92.4 | 88.2 | 91.2 |
| MobileNet + Fort Samson | 91.7 | 92.7 | 90.2 | 91.3 | 91.3 | 93.2 | 92.2 | 86.3 | 91.1 |
| DenseNet201 | 87.4 | 92.3 | 90.5 | 92.8 | 93.4 | 91.4 | 90.3 | 92.4 | **91.3** |
| DenseNet201 + Starry Night | 87.2 | 91.8 | 88.2 | 94.5 | 91.9 | 93.0 | 91.9 | 85.8 | 90.5 |
| DenseNet201 + Fort Samson | 90.5 | 92.9 | 90.5 | 91.4 | 90.7 | 93.6 | 92.6 | 83.2 | 90.7 |
| ResNet101V2 | 87.6 | 91.4 | 93.3 | 89.3 | 90.9 | 90.1 | 91.0 | 91.2 | 90.6 |
| ResNet101V2 + Starry Night | 88.8 | 91.2 | 86.5 | 92.4 | 91.1 | 93.2 | 92.9 | 82.7 | 89.9 |
| ResNet101V2 + Fort Samson | 90.3 | 89.8 | 90.1 | 92.2 | 89.8 | 93.7 | 92.8 | 88.2 | **90.9** |

**Table 2** Comparison of the popular TL methods' and the proposed techniques' $F_1$ scores and their averages on BP4D, trained with the original images and trained after both "The Starry Night" and "The View of Fort Samson" are transferred to the dataset as an augmentation method. Bolds indicate the increasing scores among each TL

| Method | 1 | 2 | 4 | 6 | 7 | 10 | 12 | 15 | 17 | 23 | 24 | Avg. |
|---|---|---|---|---|---|---|---|---|---|---|---|---|
| Xception | 76.9 | 78.9 | 78.5 | 82.7 | 77.9 | 82.4 | 89.3 | 78.7 | 75.7 | 73.8 | 82.9 | 79.8 |
| Xception + Starry Night | 76.3 | 78.3 | 80.1 | 80.5 | 79.1 | 83.3 | 85.1 | 77.1 | 70.3 | 74.8 | 80.2 | 78.6 |
| Xception + Fort Samson | 78.1 | 76.4 | 80.4 | 81.2 | 78.3 | 84.4 | 89.5 | 78.6 | 75.8 | 73.9 | 83.1 | **80.0** |
| InceptionV3 | 74.7 | 75.8 | 79.7 | 82.1 | 75.9 | 83.1 | 86.7 | 76.6 | 71.1 | 69.2 | 84.6 | 78.1 |
| InceptionV3 + Starry Night | 75.0 | 74.4 | 76.7 | 78.6 | 77.2 | 86.5 | 82.6 | 72.7 | 68.8 | 73.9 | 76.1 | 76.6 |
| InceptionV3 + Fort Samson | 75.1 | 75.8 | 79.2 | 82.4 | 78.7 | 82.1 | 83.1 | 77.4 | 72.1 | 72.2 | 84.0 | **78.4** |
| VGG16 | 76.0 | 80.2 | 73.9 | 82.4 | 79.2 | 84.1 | 86.5 | 79.1 | 76.3 | 71.1 | 82.1 | **79.2** |
| VGG16 + Starry Night | 70.0 | 72.2 | 77.6 | 78.2 | 76.1 | 84.0 | 80.5 | 76.1 | 72.1 | 72.7 | 78.5 | 76.2 |
| VGG16 + Fort Samson | 75.3 | 80.1 | 76.1 | 82.5 | 78.9 | 84.4 | 89.2 | 76.5 | 73.4 | 72.8 | 78.7 | 78.9 |
| VGG19 | 73.5 | 74.3 | 77.0 | 79.4 | 77.6 | 79.4 | 86.2 | 72.1 | 69.0 | 66.3 | 78.7 | **75.8** |
| VGG19 + Starry Night | 72.2 | 72.6 | 75.7 | 75.6 | 75.3 | 80.9 | 77.8 | 71.0 | 65.0 | 59.9 | 77.7 | 73.1 |
| VGG19 + Fort Samson | 64.3 | 76.3 | 67.9 | 77.9 | 69.5 | 79.7 | 82.7 | 72.9 | 66.9 | 68.4 | 73.3 | 72.7 |
| MobileNet | 76.4 | 74.6 | 78.9 | 79.4 | 77.9 | 80.5 | 87.3 | 73.2 | 74.0 | 76.6 | 80.6 | **78.1** |
| MobileNet + Starry Night | 74.3 | 75.8 | 73.4 | 77.8 | 69.2 | 79.9 | 82.0 | 78.1 | 67.8 | 72.2 | 79.3 | 75.4 |
| MobileNet + Fort Samson | 76.4 | 76.0 | 75.4 | 82.0 | 75.4 | 81.2 | 87.6 | 78.7 | 70.1 | 73.1 | 79.7 | 77.8 |
| DenseNet201 | 74.8 | 79.5 | 78.8 | 81.2 | 76.0 | 84.0 | 85.2 | 77.4 | 76.4 | 72.9 | 84.1 | **79.1** |
| DenseNet201 + Starry Night | 72.8 | 77.6 | 79.4 | 79.5 | 76.6 | 82.8 | 83.7 | 79.6 | 68.9 | 72.1 | 79.8 | 77.5 |
| DenseNet201 + Fort Samson | 72.7 | 75.6 | 75.1 | 78.7 | 76.4 | 81.3 | 82.3 | 74.9 | 64.0 | 74.0 | 79.7 | 75.9 |
| ResNet101V2 | 78.1 | 76.4 | 80.5 | 78.2 | 76.3 | 84.6 | 86.2 | 74.6 | 72.7 | 74.2 | 83.3 | **78.6** |
| ResNet101V2 + Starry Night | 74.4 | 75.0 | 78.0 | 77.5 | 74.1 | 82.3 | 86.2 | 75.4 | 70.7 | 70.7 | 80.2 | 76.8 |
| ResNet101V2 + Fort Samson | 78.6 | 76.8 | 78.7 | 82.1 | 76.8 | 84.2 | 87.1 | 75.6 | 71.2 | 69.1 | 82.0 | 78.4 |

- **DISFA**, despite its controlled conditions, demonstrated minimal improvements when subjected to style transfer in combination with various TL models.
- **BP4D** exhibited modest enhancements in performance on select TL models when style transfer was applied, particularly when combined with certain models.

**Table 3** Comparison of the popular TL methods' and the proposed techniques' $F_1$ scores and their averages on EmotioNet, trained with the original images and trained after both "The Starry Night" and "The View of Fort Samson" are transferred to the dataset as an augmentation method. Bolds indicate the increasing scores among each TL

| Method | 1 | 2 | 4 | 6 | 9 | 12 | 17 | 25 | 26 | Avg. |
|---|---|---|---|---|---|---|---|---|---|---|
| Xception | 65.1 | 65.3 | 69.4 | 76.2 | 76.8 | 79.3 | 70.1 | 68.6 | 57.9 | 69.9 |
| Xception + Starry Night | 66.2 | 62.5 | 69.3 | 77.3 | 75.3 | 79.6 | 67.1 | 68.6 | 60.1 | 69.6 |
| Xception + Fort Samson | 65.7 | 66.7 | 69.2 | 77.7 | 77.1 | 80.1 | 68.6 | 69.9 | 59.0 | **70.4** |
| InceptionV3 | 64.3 | 64.4 | 67.2 | 74.0 | 73.6 | 77.7 | 71.5 | 66.3 | 63.1 | 69.1 |
| InceptionV3 + Starry Night | 65.8 | 64.0 | 66.9 | 78.2 | 75.7 | 72.9 | 68.9 | 64.6 | 57.9 | 68.3 |
| InceptionV3 + Fort Samson | 66.6 | 69.2 | 67.2 | 74.5 | 75.9 | 75.2 | 71.6 | 66.7 | 65.8 | **70.3** |
| VGG16 | 67.2 | 73.2 | 66.1 | 71.9 | 75.7 | 74.4 | 63.9 | 67.5 | 62.6 | 69.2 |
| VGG16 + Starry Night | 71.8 | 67.8 | 70.1 | 78.2 | 74.4 | 73.1 | 65.7 | 67.8 | 61.9 | 70.1 |
| VGG16 + Fort Samson | 69.9 | 69.4 | 66.5 | 73.7 | 72.7 | 78.1 | 66.8 | 68.7 | 66.4 | **70.2** |
| VGG19 | 66.4 | 67.7 | 58.8 | 67.9 | 76.4 | 77.6 | 67.6 | 64.9 | 60.3 | 67.5 |
| VGG19 + Starry Night | 61.9 | 64.7 | 63.0 | 76.1 | 77.3 | 73.6 | 64.7 | 65.8 | 55.7 | 67.0 |
| VGG19 + Fort Samson | 66.6 | 69.0 | 61.2 | 71.4 | 73.4 | 76.0 | 68.4 | 65.5 | 61.2 | **68.1** |
| MobileNet | 64.5 | 68.6 | 65.3 | 71.8 | 81.5 | 76.8 | 67.6 | 63.3 | 63.4 | 69.2 |
| MobileNet + Starry Night | 64.0 | 66.1 | 64.9 | 75.0 | 79.3 | 75.9 | 64.2 | 66.4 | 59.2 | 68.3 |
| MobileNet + Fort Samson | 66.4 | 68.6 | 65.9 | 72.0 | 82.8 | 77.7 | 69.7 | 66.5 | 62.7 | **70.3** |
| DenseNet201 | 62.5 | 66.9 | 60.7 | 71.8 | 78.8 | 76.3 | 69.9 | 65.2 | 62.9 | 68.3 |
| DenseNet201 + Starry Night | 65.0 | 66.2 | 70.1 | 72.4 | 77.4 | 78.7 | 66.1 | 67.1 | 61.7 | **69.4** |
| DenseNet201 + Fort Samson | 66.3 | 62.4 | 67.6 | 73.5 | 77.4 | 76.4 | 69.1 | 65.7 | 65.2 | 69.3 |
| ResNet101V2 | 68.4 | 63.7 | 70.1 | 77.3 | 78.0 | 81.9 | 70.8 | 66.9 | 61.4 | 70.9 |
| ResNet101V2 + Starry Night | 60.1 | 65.1 | 65.1 | 73.1 | 77.9 | 79.2 | 69.7 | 66.4 | 59.0 | 68.4 |
| ResNet101V2 + Fort Samson | 68.7 | 64.1 | 70.2 | 76.1 | 77.9 | 82.7 | 69.9 | 68.9 | 62.2 | **71.2** |

- **Emotionet**, characterized by its challenging and diverse real-world scenarios, exhibited consistent improvements across a wide range of tested TL models when subjected to style transfer with the "View of Fort Samson" as the reference image.

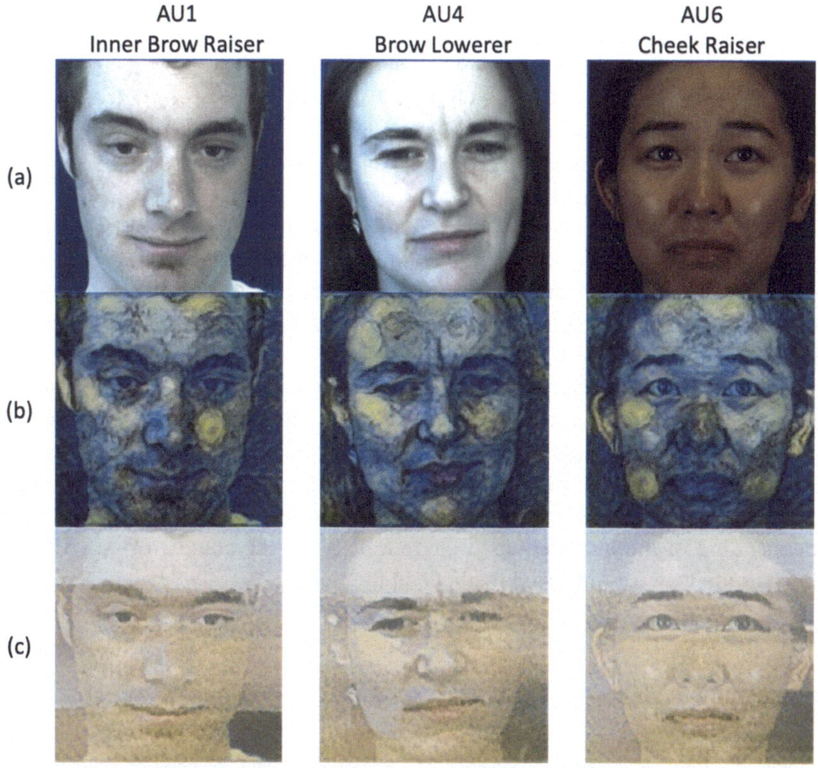


**Fig. 3** (**a**) The original image and the synthesized images after applying (**b**) the *Starry Night* and (**c**) the *View of Fort Samson*

## 4 Discussion and Conclusion

In this study, an approach for enhancing data augmentation in Facial Action Unit (AU) detection by leveraging style transfer to enhance the facial action unit dataset by synthesizing images has been proposed.

The observed trend in the experiments indicates a critical relationship between the complexity of the style-mapped image and the difficulty in achieving improved scores for facial action unit (AU) detection. In particular, as the style-transferred images become more intricate and challenging due to the inherent complexity of the artistic style (e.g., "Starry Night"), the task of AU detection becomes increasingly demanding. Minor facial movements and subtle details on the face that are crucial for AU detection can become occluded by the distortion introduced by intricate styles. This underscores the trade-off between achieving artistic style transfer and preserving the essential details for facial AU detection.

Moreover, we noticed that the dataset's complexity plays a pivotal role in the effectiveness of style transfer as an augmentation method. When dealing with challenging datasets, the application of style transfer emerges as an advantage. This is particularly evident when employing the "View of Fort Samson" style transfer, which consistently enhanced model performance across various transfer learning (TL) models. In such demanding contexts, style transfer can assist in creating more robust models for AU detection.

The results demonstrated that the effectiveness of style transfer for data augmentation is heavily influenced by the complexity of the style image and the target dataset. While more intricate styles pose challenges by distorting facial features, simpler styles prove to be more beneficial, particularly when applied to challenging datasets.

In conclusion, our findings underscore the importance of carefully considering the choice of style transfer image and the nature of the target dataset when employing style transfer as an augmentation method for AU detection. It is imperative to strike a balance between artistic style transfer and preserving the intricate details required for precise facial AU recognition. This research contributes to a deeper understanding of the implications of style transfer in the context of AU detection and paves the way for improved methods in the field.

## References


1. Umer, S., Rout, R.K., Pero, C., Nappi, M.: Facial expression recognition with trade-offs between data augmentation and deep learning features. J. Amb. Intell. Human. Comput. **13**, 1–15 (2022)
2. Garcea, F., Serra, A., Lamberti, F., Morra, L.: Data augmentation for medical imaging: a systematic literature review. Comput. Biol. Med. **152**, 106391 (2022)
3. Guo, J., Kurup, U., Shah, M.: Is it safe to drive? An overview of factors, metrics, and datasets for driveability assessment in autonomous driving. IEEE Trans. Intell. Transport. Syst. **21**(8), 3135–3151 (2019)
4. Li, W., Pan, C., Zhang, R., Ren, J., Ma, Y., Fang, J., Yan, F., Geng, Q., Huang, X., Gong, H., et al.: Aads: augmented autonomous driving simulation using data-driven algorithms. Sci. Rob. **4**(28), eaaw0863 (2019)
5. Conte, G., Zanoli, S.M., Scaradozzi, D., Caiti, A., et al.: Robotics techniques for data acquisition in underwater archeology: Int. J. Mech. Control **10**(1), 45–51 (2009)
6. Liu, J., Luo, H., Liu, H.: Deep learning-based data analytics for safety in construction. Autom. Construct. **140**, 104302 (2022)
7. Su, D., Kong, H., Qiao, Y., Sukkarieh, S.: Data augmentation for deep learning based semantic segmentation and crop-weed classification in agricultural robotics. Comp. Electron. Agricul. **190**, 106418 (2021)
8. Ronneberger, O., Fischer, P., Brox, T.: U-net: Convolutional Networks for Biomedical Image Segmentation, pp. 234–241 (2015)
9. Zhang, H., Xu, T., Li, H., Zhang, S., Wang, X., Huang, X., Metaxas, D.N.: Stackgan: Text to photo-realistic image synthesis with stacked generative adversarial networks. In: Proceedings of the IEEE International Conference on Computer Vision, pp. 5907–5915 (2017)

10. Litjens, G., Kooi, T., Bejnordi, B.E., Setio, A.A.A., Ciompi, F., Ghafoorian, M., van der Laak, J.A., van Ginneken, B., Sánchez, C.I.: A survey on deep learning in medical image analysis. Med. Image Analy. **42**, 60–88 (2017)
11. Ravi, D., Wong, C., Deligianni, F., Berthelot, M., Andreu-Perez, J., Lo, B., Yang, G.-Z.: Deep learning for health informatics. IEEE J. Biomed. Health Inf. **21**, 4–21 (2016)
12. Greenspan, H., van Ginneken, B., Summers, R.M.: Guest editorial deep learning in medical imaging: overview and future promise of an exciting new technique. IEEE Trans. Med. Imaging **35**, 1153–1159 (2016)
13. Choras, R.S.: Multimodal biometrics for person authentication. In: Security and Privacy From a Legal, Ethical, and Technical Perspective, p. 177 (2019)
14. Guo, T., Tan, X.L., DeepSecurity, G.: Lemna: Explaining deep learning based security applications. In: Proceedings of the ACM on Measurement and Analysis of Computing Systems (2018)
15. Levine, S., Popović, Z.: Learning to be a bot: Reinforcement learning in real-time strategy games. In: Proceedings of the 24th International Conference on Artificial Intelligence (2018)
16. Dixit, V.V., Chand, S., Nair, D.J.: Autonomous vehicles: disengagements, accidents and reaction times. PLoS one **11**(12), e0168054 (2016)
17. Lane, N.D., Bhattacharjee, S., Georgiev, P., Hu, H., Mongia, A., Mullen, J.: Deepx: A software accelerator for low-power deep learning inference on mobile devices. In: Proceedings of the 15th International Conference on Information Processing in Sensor Networks (2016)
18. Friesen, W.V., Ekman, P.: Facial Action Coding System: A Technique for the Measurement of Facial Movement. Consulting Psychologists Press, Washington (1978)
19. Pumarola, A., Agudo, A., Martinez, A.M., Sanfeliu, A., Moreno-Noguer, F.: Ganimation: Anatomically-aware facial animation from a single image. In: Proceedings of the European Conference on Computer Vision (ECCV), pp. 818–833 (2018)
20. Liu, Z., Song, G., Cai, J., Cham, T.-J., Zhang, J.: Conditional adversarial synthesis of 3d facial action units. Neurocomputing **355**, 200–208 (2019)
21. Pahl, J., Rieger, I., Seuss, D.: Multi-label class balancing algorithm for action unit detection. Preprint arXiv:2002.03238 (2020)
22. Mavadati, S.M., Mahoor, M.H., Bartlett, K., Trinh, P., Cohn, J.F.: Disfa: a spontaneous facial action intensity database. IEEE Trans. Affect. Comput. **4**(2), 151–160 (2013)
23. Zhang, X., Yin, L., Cohn, J.F., Canavan, S., Reale, M., Horowitz, A., Liu, P., Girard, J.M.: Bp4d-spontaneous: a high-resolution spontaneous 3d dynamic facial expression database. Image Vision Comput. **32**(10), 692–706 (2014)
24. Fabian Benitez-Quiroz, C., Srinivasan, R., Martinez, A.M.: Emotionet: An accurate, real-time algorithm for the automatic annotation of a million facial expressions in the wild. In: Proceedings of the IEEE Conference on Computer Vision and Pattern Recognition, pp. 5562–5570 (2016)
25. Cakir, D., Arica, N.: Cascading cnns for facial action unit detection. Eng. Sci. Technol. Int. J. **47**, 101553 (2023)
26. Viola, P., Jones, M.J.: Robust real-time face detection. Int. J. Comput. Vision **57**, 137–154 (2004)
27. Deng, J., Dong, W., Socher, R., Li, L.-J., Li, K., Fei-Fei, L.: Imagenet: A large-scale hierarchical image database. In: 2009 IEEE Conference on Computer Vision and Pattern Recognition, pp. 248–255. IEEE, Piscataway (2009)
28. Kasap, O.Y., Cakir, D.: Analysis of transfer learning models for face mask detection. In: 2022 IEEE 12th Annual Computing and Communication Workshop and Conference (CCWC), pp. 0199–0202. IEEE, Piscataway (2022)

# Best Image Processing for Higher Face Detection Rate Using Haar Cascades

**Suneet Gupta, Uday Singh, Manoj Rawat, Asha Kurian, Lopa Mandal, Praveen Gupta, and Tanvir Habib Sardar**

## 1 Introduction

Face detection is an important topic in image processing and specifically computer vision. It is the primary and crucial step of face recognition. It has an advantage over other biometrics systems like fingerprints and palm prints as it can be carried out without contact with the person. Face detection and recognition have obviously helped the development of human–computer interaction. Yang et al. did a survey and presented the trends of research in face detection [1]. One common problem in face detection algorithms, including the Haar face detection algorithm as mentioned by Viola and Jones [2], is that the false positives cannot be avoided. False positive is a situation where the algorithm treats nonfacial region as a facial region. In other words, appropriate hit rates, that is, true positives, and low missing detections, that is, false negatives, are hard tasks in face detection as mentioned by Galbally et al. [3]. Zhang et al. [4] cited that the issues related to rotation, pose, and complicated background give rise to a number of false positives. Face analytics basically deals with extracting information like landmarks, expression, gender, and age identity. Some of the main application areas are law enforcement, self-driving vehicles, authentication for banking and security system access, and face biometrics for payments. Face recognition is a significant and integral part of today's emerging *biometrics* and *video surveillance* markets. Generally, there are three modules of a face identification and verification system. As stated in [5, 6], the first one is face detection, which is used for localizing faces. Refs. [5, 6] mentioned that the second

S. Gupta (✉) · U. Singh · M. Rawat · A. Kurian · L. Mandal · P. Gupta
CSE Department, Alliance University, Bangalore, India

T. H. Sardar
CSE Department, GITAM University, Bangalore, India



Ö. M. Gül et al. (eds.), *7th EAI International Conference on Robotic Sensor Networks*, EAI/Springer Innovations in Communication and Computing,
https://doi.org/10.1007/978-3-031-64495-5_9

module is used to localize landmarks such as eye centers, tip of the nose, corners of the mouth, and tips of ear lobes. These landmarks are used to align faces, which alleviates the effects of in-plane rotation and scale. The third and final module, as also pointed out by Bansal et al. [7], is a feature extractor that includes all the information of identity details in a descriptor of high dimension.

A new and improvised algorithm regarding face detection and recognition system was presented by Idelette et al. [8]. It combines KNN with LBP to resolve face detection issues like pose, illumination, and expression. In one instance, Priadana and Habibi showed Haar feature-based face detection combined with web detection technique was used to filter selfie face images on Instagram [9]. The greatest hurdles in designing a successful face detection and recognition system are scale, illumination, pose, occlusion, background, and expression. Several algorithms have evolved keeping in mind these challenges. Ramaiah et al. [10] published their work on issues related to illumination using convolutional neural networks (CNNs). CNNs have proven to be very effective for numerous computer vision jobs like object detection as suggested in works in [11, 12] and image classification as suggested in works in [4, 13]. As far as face recognition is concerned, not only single-face but multiface recognition processes have been developed by Mantoro et al. [14]. Some other recent works related to face detection and recognition using the Haar classifier and LBP histogram were illustrated by Khan et al. [15]. Haar classifier can detect any object, the only thing being that it must be trained to identify that object. With the latest technologies of computing, high-end processors, and such decreased cost of video image acquisition, it has been rightly pointed out by Yang et al. [16] that as per the works in [17–19], computer vision systems can be deployed in desktop and embedded systems.

Today, we have many commercially available face processing systems. The first and foremost step of these face processing systems is face detection, whose efficiency decides the efficiency of the subsequent steps. Detecting faces from an image is a daunting task because of the challenging issues related to it. These issues are variability in scale, pose (frontal, 45-degree, profile, upside down), structural components (beards, mustaches, and glasses), location, and orientation (upright, rotated). The other factors that change the overall appearance of a face are facial expressions (happy, sad, disgusted, etc.), occlusion (some faces may partially occlude other faces), and lighting (spectra, source distribution, and intensity). Several techniques of face detection that are available at present are feature-based face detection, geometric-based face detection, neural network-based face detection, example-based face detection, and many more as cited by Sung and Poggio [20] and Atharifard and Ghofrani [21]. According to Yang et al. [16], the definition of face detection is as follows: Given an arbitrary image, the goal of face detection is to determine whether there are any faces in the image and, if present, return the image location and extent of each face. Recently, face detection methods using learning algorithms have sought the attention of researchers and outperformed the methods based on other approaches.

The performance of a face detection algorithm can be evaluated by taking the ratio between the number of faces detected correctly and the total number of faces

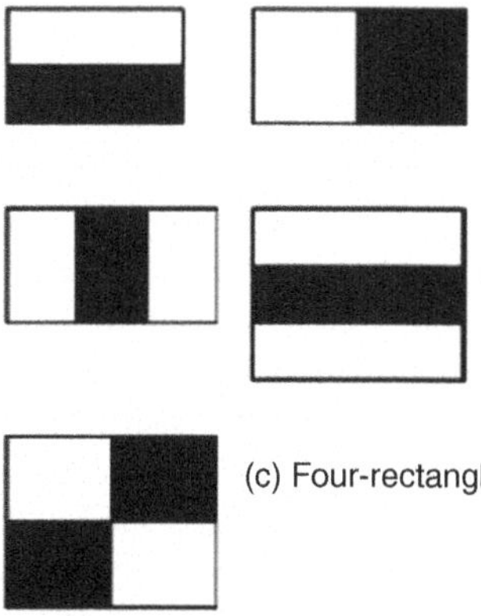


**Fig. 1** Haar features

as counted by a human. This ratio is called the detection rate. Image region is said to be correctly detected as a face if it contains more than a certain percentage of face within itself. Generally, face detection algorithms make two types of errors: false negatives and false positives. False negative is where faces are missed, and we get a low detection rate. False positive is where the detector claims an image region to be a face, but it is not.

## 2 Face Detection Using Haar Cascades

The chapter intends to improve the performance of face detection using the Haar feature-based cascade classifiers. Viola and Jones [2] proposed an effective method of object detection using the Haar cascade classifiers. The method is based on learning concepts in which a cascade function is trained with positive and negative images. Then, it is used to detect objects in other images. The objects in our case are faces, and to train the classifier many positive and negative images are needed. Positive images are those with faces and negative images are those without faces. Features are extracted using the Haar features as shown in Fig. 1. They behave just like convolutional kernel. Each feature value is obtained by subtracting the sum of pixel values under the white region from the sum of the pixel values under the black region those with faces and the negative images are those without faces.

## 3 Finding the Best Image Processing Technique

This chapter finds out the best image processing technique, which when used as a preprocessing increases the detection rate of the Haar feature-based face detection algorithm. For this purpose, two sets, each of 15 images, were taken from different sources on the Internet. Let these sets be called *analysis part* and *testing part.* Ten images of *testing part* were taken from the WIDER FACE dataset by Yang et al. [1].

The method consists of applying different preprocessing techniques to the images of analysis part and checking the result of implementation of the Haar algorithm on them. Normally a face detection algorithm is implemented on the grayscale version of the original image. Therefore, it is the grayscale image that will be preprocessed differently and the result of the implementation of the Haar algorithm on each preprocessed image will be checked. These preprocessed images can also be called test images. When the Haar algorithm is implemented on test images, the result is checked to determine whether it has detected any new faces (true positives) compared to the faces detected with the image that did not undergo any preprocessing, that is, the grayscale version. Altogether, there are 32 preprocessing techniques or test images considered, including the one without any preprocessing. The preprocessing techniques considered are the ones that basically enhance the image in some way or the other keeping the edges unaffected to the maximum extent. The resulting details of the implementation of the Haar face detection algorithm on test images are shown in Table 1. For example, the resulting details of the implementation of the Haar algorithm on the gray image version of the first image (image p1) of the formulating part are entered in the first row (s. no. 1) of Table 1. The second row (s. no. 2) of Table 1 contains the resulting details when the grayscale image of image p1 was preprocessed with the histogram equalization process prior to the implementation of the Haar algorithm. Similarly, the resulting details of other test images of image p1 are shown in Table 1.

## *3.1 Description of the Column Heads in Table 1*

(a) Len (faces): Total number of rectangles/windows drawn in an attempt to detect faces.
(b) TP: True positive, number of rectangles that properly contained faces.
(c) FP: False positive, number of rectangles not properly or not at all contained faces.
(d) Repeat TP: True positives, which were already present in the normal implementation, that is, using gray image (Gray_Img).
(e) New TP: True positives, which were not present in the normal implementation, that is, using gray image (Gray_Img).
(f) Test image: Explainedin the following section separately as it is needs a detailed explanation.

**Table 1** Face detection with different test images of the image p1

| | Group photo ➔ | Image p1 (#faces = 30) | | | | |
|---|---|---|---|---|---|---|
| S. no. | Test image | Len (faces) | TP | FP | Repeat TP | New TP |
| 1. | Gray_Img | 20 | 19 | 1 | – | |
| 2. | H_Eq_Gray_Img | 19 | 18 | 1 | 17 | KLR |
| 3. | Lap_Fil | 0 | 0 | 0 | 0 | |
| 4. | H_Eq_Lap_Fil | 2 | 2 | 0 | 2 | |
| 5. | Lap_Img | 20 | 19 | 1 | 18 | MSD |
| 6. | H_Eq_Lap_Img | 20 | 19 | 1 | 17 | KLR,Coach |
| 7. | Sob_Fil (scharr) | 0 | 0 | 0 | 0 | |
| 8. | H_Eq_Sob_Fil | 3 | 1 | 2 | 0 | KLR |
| 9. | Sob_Img (scharr) | 11 | 11 | 0 | 11 | |
| 10. | H_Eq_Sob_Img | 16 | 15 | 1 | 15 | |
| 11. | Mask_Img | 13 | 13 | 0 | 13 | |
| 12. | H_Eq_Mask_Img | 18 | 17 | 1 | 16 | KLR |
| 13. | Sob_Lap | 15 | 15 | 0 | 14 | KLR |
| 14. | H_Eq_Sob_Lap | 18 | 17 | 1 | 17 | |
| 15. | Med_Fil | 17 | 17 | 0 | 16 | KLR |
| 16. | H_Eq_Med_Fil | 18 | 17 | 1 | 16 | KLR |
| 17. | Med_Img | 17 | 17 | 0 | 16 | KLR |
| 18. | H_Eq_Med_Img | 18 | 17 | 1 | 16 | KLR |
| 19. | UnSh_Img | 19 | 19 | 0 | 18 | MSD |
| 20. | H_Eq_UnSh_Img | 18 | 18 | 0 | 17 | KLR |
| 21. | H_Boost_Fil | 20 | 20 | 0 | 18 | MSD,KLR |
| 22. | H_Eq_ H_Boost_Fil | 20 | 19 | 1 | 18 | KLR |
| 23. | H_Boost_Img | 20 | 20 | 0 | 18 | MSD,KLR |
| 24. | H_Eq_ H_Boost_Img | 20 | 19 | 1 | 18 | KLR |
| 25. | Bil_Fil | 16 | 16 | 0 | 16 | |
| 26. | H_Eq_Bil_Fil | 18 | 17 | 1 | 16 | KLR |
| 27. | Bil_Img | 16 | 16 | 0 | 16 | |
| 28. | H_Eq_Bil_Img | 18 | 17 | 1 | 16 | KLR |
| 29. | Bil_Fil_ConAd | 16 | 16 | 0 | 16 | |
| 30. | H_Eq_Bil_Fil_ConAd | 18 | 17 | 1 | 16 | KLR |
| 31. | Bil_Img_ConAd | 17 | 17 | 0 | 16 | KLR |
| 32. | H_Eq_Bil_Img_ConAd | 17 | 16 | 1 | 16 | |

## *3.2 Description of Test Images*

The following were the different versions of test images taken for implementing the Haar face detection cascades.

1. *Gray_Img*

Grayscale converted image of the original image. The image that is taken for face detection is first converted into the grayscale image, then the Haar face detection

**Fig. 2** Faces detected on Gray_Img

classifiers are applied to it. This is how face detection is normally/usually done. This is the test image without any preprocessing. It can be seen in Fig. 2 that with this test image, out of 30 faces, 19 were detected and 1 FP was also obtained. The same has been entered in Table 1.

2. *H_Eq_Gray_Img*

Histogram is nothing but the intensity distribution of the image. It shows us the number of pixels for each intensity level. For a digital image with intensity levels in the range of [0, L-1], it is a discrete function and is expressed as

$$h\left(r_{\mathrm{k}}\right)=n_{\mathrm{k}} \tag{1}$$

where $r_{\mathrm{k}}$ is the kth intensity level and $\mathrm{n_k}$ is the number of pixels with intensity $\mathrm{r_k}$

$$p\left(r_{\mathrm{k}}\right)=\frac{n_{\mathrm{k}}}{\mathrm{MXN}} \tag{2}$$

where MXN is the size of the image and $p(r_{\mathrm{k}})$ is the probability of a pixel having intensity $r_{\mathrm{k}}$.

Histogram equalization modifies the intensity distribution of the image such that it gives a linear trend to the cumulative probability function. It tries to distribute the number of pixels uniformly among the number of intensity levels, producing a flat histogram with a high contrast. This transformation, $T(r_{\mathrm{k}})$, was given by Gonzalez and Woods [22] and can be written as

**Fig. 3** Faces detected on H_Eq_Gray_Img

$$s_k = T\,(r_k) = (L - 1)\sum_{j=0}^{k} \boldsymbol{p}\,(\boldsymbol{r}_k) \tag{3}$$

where $k = 0{,}1{,}2\ldots.$L-1.

It can be seen in Fig. 3 that 18 faces (TPs) were detected and 1 FP was also there. One important point to note here is that out of the 18 TPs 17 were also detected in Fig. 2. One TP (KLR, i.e., KL Rahul, cricketer) that is detected here was not detected in Fig. 2. So, this new TP is shown in Fig. 3.

3. *Lap_Fil*

Laplacian is the second derivative operator.

Laplacian gives us an isotropic measure of rapid intensity changes in an image. The two commonly used kernels are shown in Fig. 4. It is equally valid to use the signs the other way round, the central pixel being negative and the others being positive. The Laplacian can be calculated using the standard convolution method by any one kernel shown in Fig. 4. Here, the right-side kernel was used. No faces were detected, and hence, all the entries in the third row of Table 1 are zeros.

Note that the resulting images with detected faces henceforth are not shown to keep the length of the manuscript minimum. However, the resulting details are noted and shown in Table 1.

4. *H_Eq_Lap_Fil*

Histogram equalized image of Lap_Fil.

**Fig. 4** Two common Laplacian

| | | |
|---|---|---|
| 0 | −1 | 0 |
| −1 | 4 | −1 |
| 0 | −1 | 0 |

| | | |
|---|---|---|
| −1 | −1 | −1 |
| −1 | 8 | −1 |
| −1 | −1 | −1 |

**Fig. 5** Sobel filter

| | | |
|---|---|---|
| -1 | 0 | 1 |
| -2 | 0 | 2 |
| -1 | 0 | 1 |

$G_x$

| | | |
|---|---|---|
| -1 | -2 | -1 |
| 0 | 0 | 0 |
| -1 | 0 | 1 |

$G_y$

**Fig. 6** Scharr operators

| | | |
|---|---|---|
| -3 | 0 | 3 |
| -10 | 0 | 10 |
| -3 | 0 | 3 |

$G_x$

| | | |
|---|---|---|
| -3 | -10 | -3 |
| 0 | 0 | 0 |
| 3 | 10 | 3 |

$G_y$

5. *Lap_Img*

Laplacian filtered image of the Gray_Img. Gray_Img was added to the filtered image.

The new TP is MSD (MS Dhoni, cricketer). MSD was not detected in Gray_Img. One FP was there, and the remaining 18 were repeated, that is, they were also detected in Gray_Img.

6. *H_Eq_Lap_Img*

Histogram equalized image of Lap_Img. KLR and coach are the new faces detected here.

7. *Sob_Fil (Scharr)*

Sobel filtered image of Gray_Img.

The Sobel operator is a first-order discrete differentiation operator. It approximates the gradient of the intensity function of an image. The Sobel filters are shown in Fig. 5. Actually, the Sobel filter of size 3 × 3 shows some inaccuracies, and in OpenCV, we also have another function, Scharr ( ) function [23], which takes care of the inaccuracies in Sobel. Therefore, in this chapter, Scharr filters are used. The Scharr operators are shown in Fig. 6.

8. *H_Eq_Sob_Fil*

Histogram equalized image of Sob_Fil.

9. *Sob_Img (Scharr)*

Sobel (Scharr actually) filtered image of Gray_Img. Gray_Img was added to the filtered image.

10. *H_Eq_Sob_Img*

Histogram equalized image of Sob_Img.

11. *Mask_Img*

Mask_Img is obtained by multiplying Laplacian with Sobel image as shown in Gonzalez and Woods [22]. Here, Laplacian is multiplied with smoothed Sobel image because being a second-order derivative it is highly sensitive to noise.

Both Laplacian and gradient operator are used to highlight intensity discontinuities. Laplacian is better for enhancing fine details and noise while gradient works better if edges are to be enhanced. Therefore, the idea of multiplying both these images, that is, Laplacian and Sobel gradient, gives us an image in which both fine details and edges are highlighted.

12. *H_Eq_Mask_Img*

Histogram equalized image of Mask_Img.

13. *Sob_Lap*

Sum of Lap_Img and Sob_Img. Multiplication of Laplacian and Sobel gradient has already been tested; let us also see the effect of adding them.

14. *H_Eq_Sob_Lap*

Histogram equalized image of Sob_Lap.

15. *Med_Fil:*

Median filtered image of Gray_Img. Median filtering is a nonlinear method used to remove noise from images. The reason for considering this filter is that it is very effective at removing noise while preserving edges. It is best at removing "salt and pepper"-type noise. The median filter works by sliding through the image pixel by pixel, replacing the current location value with the median value of neighboring pixels, that is, the pixel values encompassed by the filter. The median is calculated by sorting the pixel values and selecting the middle pixel value.

16. *H_Eq_Med_Fil*

Histogram equalized image of Med_Fil.

17. *Med_Img*

Median filtered image of Gray_Img. Gray_Img was added to the filtered image.

18. *H_Eq_Med_Img*

Histogram equalized image of Med_Img.

19. *UnSh_Img*

For this version of test image, let us discuss the following:

Unsharp Masking and High Boost Filtering [24]

This is a process where we use a smoothing filter for sharpening and is done as follows:

(a) The first step is to blur the image. In this chapter, Gaussian blur was used. By blurring/smoothing an image, the high-frequency components are suppressed.
(b) Then, the smoothed image is subtracted from the original image. The resultant image contains the high-frequency components, and let us call it a mask.
(c) Adding the mask to the original image enhances the high-frequency components.

Since the mask is created using a blurred/unsharp image, it is called unsharp masking. Thus, in unsharp masking, we can get the mask m(x,y) as

$$m\,(x,y) = f\,(x,y) - f_b\,(x,y) \tag{4}$$

where $f(x,y)$ is the original image and $f_b(x,y)$ is the blurred/unsharped version of the original image. Then the sharpened image can be obtained as

$$g\,(x,y) = f\,(x,y) + k^{*}\,m\,(x,y) \tag{5}$$

where k stipulates the portion of mask to be added. When $k = 1$, this is unsharp masking. When $k > 1$, this is high-boost filtering because the high-frequency components are boosted by attaching more weight to the masked (edge) image. The above two equations can be combined into one as the weighted average of the original and the blurred image:

$$g\,(x,y) = (k+1)^{*}\,f\,(x,y) - k^{*}\,f_b\,(x,y) \tag{6}$$

20. *H_Eq_UnSh_Img*

Histogram equalized image of UnSh_Img.

21. *H_Boost_Fil*

In (5) or (6), if $k>1$, then it is high-boost filtering. Here, the value of k was taken to be equal to 2.

22. *H_Eq_H_Boost_Fil*

Histogram equalized image of H_Boost_Fil.

23. *H_Boost_Img*

High boost filtered image of Gray_Img. Gray_Img was added to the filtered image.

24. *H_Eq_H_Boost_Img*

Histogram equalized image of H_Boost_Img.

25. *Bil_Fil*

Bilateral filtered image of Gray_Img.

Bilateral filter is a special type of nonlinear filter that removes noise, keeping the edges preserved. The most popular noise removal method is Gaussian blurring but it does that at the cost of blurring the edges. The bilateral filter keeps in mind the structure of the image, behaves like a classical blurring filter on areas of images without edges, and changes its behavior when it encounters an edge and does not blur the edge. The bilateral filter [25] does not only take into account the Gaussian filter, which is a function of space (space weight), but also a Gaussian function of intensity difference (range weight) that ensures only those pixels are considered for blurring that have a similar intensity as the central pixel. Gaussian blurring can be computed as follows:

$$\mathrm{GB}[I]_p = \frac{1}{W_p} \sum_{q \in S} G_\sigma \left(\|p - q\|\right) I_q \tag{7}$$

Here, $\mathrm{GB[I]_p}$ is the result at pixel *p*, and it is calculated by taking the sum over all pixels *q* weighted by the Gaussian function. $I_q$ is the intensity at pixel *q*. The bilateral filter can be computed as follows:

$$BF[I]_p = \frac{1}{W_p} \Sigma_{q \in S}\, G_{\sigma_s}\left(\|p - q\|\right) G_{\sigma_r}\left(I_p - I_q\right) I_q \tag{8}$$

**Normalization Factor** — **Space Weight** — **Range Weight**

where $W_p = \sum_{q \in S} G_{\sigma_s}\left(\|p - q\|\right) G_{\sigma_r}\left(I_p - I_q\right)$;

$G_{\sigma_s}$ is a spatial Gaussian
$G_{\sigma_r}$ is a range Gaussian
$\sigma_s$ is the size of neighborhood/kernel
$\sigma_r$ refers to the minimum amplitude of an edge

The lesser the value of $\sigma_r$, the more is the sharpness of the edge. As the value of $\sigma_r$ tends to infinity, the equation tends to Gaussian blur.

26. *H_Eq_Bil_Fil*

Histogram equalized image of Bil_Fil.

27. *Bil_Img*

Bilateral filtered image of Gray_Img. Gray_Img was added to the filtered image.

28. *H_Eq_Bil_Img*

Histogram equalized image of Bil_Img.

29. *Bil_Fil_ConAd*

This was taken from Bah and Ming [26]. The authors first adjusted the contrast of the input image by using (9). They tested with different values of $\alpha$ and $\beta$ and found that their face recognition algorithm worked best with $\alpha = 1.5$ and $\beta = 0.0$. That is why we also went ahead with the same values. The expression used for obtaining contrast-adjusted gray image is

$$\text{Gray\_Img\_ConAd} = \alpha^* f(x, y) + \beta \quad (9)$$

After the contrast adjustment of the input image, the bilateral filter was implemented.

30. *H_Eq_Bil_Fil_ConAd*

Histogram equalized image of Bil_Fil_ConAd.

31. *Bil_Img_ConAd*

Bilateral filtered image of the Gray_Img_ConAd. Gray_Img_ConAd was the same as obtained using (9). Gray_Img_ConAd was added to the filtered image, that is, Bil_Img_ConAd was obtained by adding Bil_Fil_ConAd and Gray_Img_ConAd.

32. *H_Eq_Bil_Img_ConAd*

Histogram equalized image of Bil_Img_ConAd.

The images or group photos for the analysis part were not just randomly taken from the Internet, but the issues that act as hindrances in making an efficient face processing system were also considered. As pointed out earlier, the issues were mainly pose, orientation, facial expressions, lighting conditions, etc.

## 4 Proposed Method

Now that we are done with the analysis, we will select the most effective test image. Similar to how Table 1 was created for image p1, similar tables for the 14 remaining images or group photos of the analysis part were also created. The "New TP" columns of all the 15 tables were considered, and for each of the test images, the sum of the total number of new TPs was calculated. The maximum number of new TPs was yielded by the test image H_Eq_Lap_Img, which was found to be equal to 15. So, the method of implementing the Haar algorithm is to implement first on the gray image and note the number of TPs. The second step is to take the H_Eq_Lap_Img version of the image, implement the Haar algorithm again, and note the number of new TPs. Both TPs are then added to get a higher detection rate.

## 5 Testing and Results

This is the testing phase of the proposed method. For this, one more set of 15 images was created. Let this set be called the testing part. Out of the 15 images, 10 were taken from Yang et al. [2] The WIDER FACE dataset and 5 were taken from the Internet. The latter five images were challenging as they had upside-down faces, low-lighting issues, face with turban, etc. The detection rate increased from 50% to 100%.

The result of the proposed method is compared side by side with the result of normal implementation of the Haar feature-based face detection algorithm. The details of the comparisons between the usual and proposed methods are shown in Table 2. The performances of the methods on images belonging to the testing part clearly show that the proposed method showed better results than the normal way of implementing the Haar feature-based face detection algorithm.

## 6 Conclusion

As mentioned earlier, detection rate is the ratio between the number of faces correctly detected and the total number of faces actually present in the image. As shown in Table 2, by using the proposed method the detection rate has increased from 8.9% in the case of *0_Parade_marchingband_1_47* to 50% in cases of *10_People_Marching_People_Marching_2_243* and *Testing part image 13*. This does not mean that the detection rate with the proposed method will always fall in this range, 8.9–50%. The chances of no increase in the detection rate are also there, but those cases will be rare. The chances of an increase in detection rate going beyond 50% are also there, but those cases will be less. In general, it can be said that with the proposed method there is a high chance of the detection rate being increased. Like in the usual method, the FPs cannot be avoided in the proposed method too.

**Table 2** Performance comparison

| S. no. | Image (testing part) | No. of TPs | | Total no. of faces | Detection rate | | Increase in detection rate |
|---|---|---|---|---|---|---|---|
| | | Normal implementation | Proposed method | | Normal implementation | Proposed method | |
| 1. | 0_Parade_marchingband_1_47 | 9 | 14 | 56 | 9/56 | 14/56 | 8.9% |
| 2. | 3_Riot_Riot_3_593 | 15 | 18 | 21 | 15/21 | 18/21 | 14.3% |
| 3. | 20_Family_Group_Family_Group_20_1031 | 2 | 4 | 5 | 2/5 | 4/5 | 40% |
| 4. | 10_People_Marching_People_Marching_2_243 | 1 | 2 | 2 | ½ | 2/2 | 50% |
| 5. | 20_Family_Group_Family_Group_20_543 | 1 | 2 | 9 | 1/9 | 2/9 | 11.1% |
| 6. | 9_Press_Conference_Press_Conference_9_75 | 16 | 19 | 21 | 16/21 | 19/21 | 14.3% |
| 7. | 8_Election_Campain_Election_Campaign_8_532 | 7 | 9 | 10 | 7/10 | 9/10 | 20% |
| 8. | 9_Press_Conference_Press_Conference_9_711 | 2 | 3 | 3 | 2/3 | 3/3 | 33.3% |
| 9. | 20_Family_Group_Family_Group_20_626 | 11 | 13 | 16 | 1/16 | 13/16 | 12.5% |
| 10. | 9_Press_Conference_Press_Conference_9_96 | 6 | 8 | 8 | 6/8 | 8/8 | 25% |
| 11. | Testing part image 11 | 4 | 5 | 6 | 4/6 | 5/6 | 16.67% |
| 12. | Testing part image 12 | 7 | 11 | 15 | 7/15 | 11/15 | 26.67% |
| 13. | Testing part image 13 | 6 | 12 | 12 | 6/12 | 12/12 | 50% |
| 14. | Testing-part image 14 | 2 | 3 | 5 | 2/5 | 3/5 | 20% |
| 15. | Testing part image 15 | 2 | 3 | 7 | 2/7 | 3/7 | 14.29% |

## References

1. Yang, S., Luo, P., Loy, C.C., Tang, X.: Wider face: a face detection benchmark. In: Proceedings of the IEEE Conference on Computer Vision and Pattern Recognition, pp. 5525–5533 (2016)
2. Viola, P., Jones, M.: Rapid object detectionusing a boosted cascade of simple features. IEEE Conf. Comput. Vis. Pattern Recognit. **1**, I-511 (2001). https://doi.org/10.1109/CVPR.2001.990517
3. Galbally, J., Marcel, S., Fierrez, J.: Biometric antispoofing methods: a survey in face recognition. IEEE Access. **2**, 1530–1552 (2014)
4. He, K., Zhang, X., Ren, S., Sun, J.: Deep residual learning for image recognition. In: IEEE Conference on Computer Vision and Pattern Recognition (CVPR), pp. 770–778 (2016)
5. Ranjan, R., Sankaranarayanan, S., Castillo, C.D., Chellappa, R.: An all-in-one convolutional neural network for face analysis. In: IEEE International Conference on Automatic Face and Gesture Recognition (FG) (2017)
6. Ranjan, R., Patel, V., Chellappa, R.: Hyperface: a deep multitask learning framework for face detection, landmark localization, pose estimation, and gender recognition. In: IEEE Transactions on Pattern Analysis and Machine Intelligence (2017)
7. Bansal, A., Ranjan, R., Castillo, C.D., Chellapp, R.: Deep features for recognizing disguised faces in the wild. In: Proceedings of IEEE Conference on Computer Vision and Pattern Recognition Workshops, pp. 10–16 (2018)
8. Idelette, B., Kambi, L., Chunsheng, G.: Enhancing face identification using local binary patterns and K-nearest neighbors. J. Imaging. **3**(37), 1–12 (2017)
9. Priadana, A., Habibi, M.: Face detection using haar cascades to filter selfie face image on instagram. In: International Conference of Artificial Intelligence and Information Technology (ICAIIT), pp. 6–9 (2019). https://doi.org/10.1109/ICAIIT.2019.8834526
10. Ramaiah, N.P., Ijjina, E.P., Mohan, C.K.: Illumination invariant face recognition using convolutional neural networks. In: IEEE International Conference on Signal Processing, Informatics, Communication and Energy Systems (SPICES), Kozhikode, India, pp. 1–4 (2015). https://doi.org/10.1109/SPICES.2015.7091490
11. Ren, S., He, K., Girshick, R., Sun, J.: Faster r-cnn: Towards real-time object detection with region proposal networks. In: Advances in Neural Information Processing Systems, pp. 91–99 (2017)
12. Liu, W., Anguelov, D., Erhan, D., Szegedy, C., Reed, S., Fu, C.-Y., Berg, A.C.: Ssd: Single shot multibox detector. In: European Conference on Computer Vision (ECCV), pp. 21–37 (2016)
13. Szegedy, C., Liu, W., Jia, Y., Sermanet, P., Reed, S., Anguelov, D., Erhan, D., Vanhoucke, V., Rabinovich, A.: Going deeper with convolutions. arXiv preprint arXiv, 1409.4842 (2015)
14. Mantoro, T., Ayu, M.A., Suhendi: Multi-faces recognition process using Haar cascades and Eigenface methods. In: 6th International Conference on Multimedia Computing and Systems (ICMCS), pp. 1–5 (2018). https://doi.org/10.1109/ICMCS.2018.8525935
15. Khan, M., Chakraborty, S., Astya, R., Khepra, S.: Face detection and recognition using OpenCV. In: International Conference on Computing, Communication, and Intelligent Systems (ICCCIS), pp. 116–119 (2019). https://doi.org/10.1109/ICCCIS48478.2019.8974493
16. Yang, M.H., Kriegman, D.J., Ahuja, N.: Detecting faces in images: a survey. IEEE Trans. Pattern Anal. Mach. Intell. **24**(1), 34–58 (2002)
17. Pentland, A.: Looking at people. IEEE Trans. Pattern Anal. Mach. Intell. **22**(1), 107–119 (2000)
18. Pentland, A.: Perceptual intelligence. Comm. ACM. **43**(3), 35–44 (2000)
19. Pentland, A., Choudhury, T.: Face recognition for smart environments. Computer. **33**(2), 50–55 (2000). https://doi.org/10.1109/2.820039
20. Sung, K., Poggio, T.: Example-based learning for view-based human face detection. IEEE Trans. Pattern Anal. Mach. Intell. **20**(1), 39–51 (1998). https://doi.org/10.1109/34.655648
21. Atharifard, A., Ghofrani, S.: Robust component-based face detection using color feature. In: Proceedings of the World Congress on Engineering, vol. II (2011)

22. Gonzalez, R., Woods, R.: Image enhancement in the spatial domain. In: Digital Image Processing, 3rd edn. Prentice Hall, Upper Saddle River (2006)
23. https://docs.opencv.org/3.4/d2/d2c/tutorial_sobel_derivatives.html
24. https://theailearner.com/2019/05/14/unsharp masking-and-highboost- filtering
25. https://www.geeksforgeeks.org/python-bilateral-filtering
26. Bah, S.M., Ming, F.: An improved face recognition algorithm and its application in attendance management system. Array. **5**, 100014., www.elsevier.com/journals/array/2590-0056/open-access-journal (2020)

# Prediction of Road Traffic Accident Severity Using Machine Learning Techniques in the Case of Addis Ababa

**Betelhem Zewdu Wubineh, Yigezu Agonafir Asamenew, and Semachew Molla Kassa**

## 1 Introduction

A road traffic accident (RTA) is any vehicle accident that takes place on a public road, whether it starts, ends, or involves a vehicle partially on the highway [1]. These accidents include collisions between vehicles and moving objects, such as fixed obstructions, pedestrians, or animals. It is also known as a motor vehicle collision (MVC) and occurs when a car collides with another car, on the side of the road, or with another immovable object. The result of the accident is significant damage to the victims, their families, and entire countries in terms of physical damage, death, and property damage [2]. Bicyclists, motorcycle riders, and pedestrians are included in all fatalities and injuries [3]. It is one of the major problems in the world today that causes deaths and financial harm [4]. Road accident injury is one of the leading causes of death in children and young adults aged 5–29 years [5]. As a result of a car accident, 1.3 million people were killed and between 20 and 50 million people suffered nonfatal injuries, many of them suffering a disability [6].

B. Z. Wubineh (✉)
Faculty of Information and Communication Technology, Wroclaw University of Science and Technology, Wroclaw, Poland

College of Engineering and Technology, Wachemo University, Hosaena, Ethiopia
e-mail: betelhem.wubineh@pwr.edu.pl

Y. A. Asamenew
Department of Information Technology, College of Informatics, University of Gondar, Gondar, Ethiopia

S. M. Kassa
College of Engineering and Technology, Wachemo University, Hosaena, Ethiopia

Bahir Dar Institute of Technology, Faculty of Civil and Water Resources Engineering, Bahir Dar University, Bahir Dar, Ethiopia

 
Ö. M. Gül et al. (eds.), *7th EAI International Conference on Robotic Sensor Networks*, EAI/Springer Innovations in Communication and Computing,
https://doi.org/10.1007/978-3-031-64495-5_10

Road crashes in developing countries are more than twice as common as in developed countries, at a rate of 13.4 and 32.2 per 100,000 people in Europe and Africa, respectively [7]. In low-income countries such as Africa, the majority of accidents on roads occur among walkers, passengers, motorized two-wheeler riders, and occupants of buses and minibuses [8]. Compared with high-income countries, people in low income countries are more likely to be involved in traffic accidents [5]. Unlicensed drivers from households with lower socioeconomic status demonstrated elevated relative risks of road accidents in the range of 1.75–3.25 [9]. Ethiopia has the highest accident rate because most transportation is by road. According to the World Health Organization (WHO) 2020 report, road traffic accidents killed 31,564 people in Ethiopia, representing 5.6% of all deaths. The age-adjusted death rate of 42.41 per 100,000 population ranks Ethiopia #19 in the world [10].

In addition, road accidents are the sixth leading cause of death in Ethiopia. The underlying research problem that initiated this research is that road traffic accidents are the leading cause of deaths and injuries at various levels in Ethiopia. Since most of the transportation is by road, a significant number of accidents occurred in Ethiopia. Research has indicated that millions of people lose their lives each year due to accidents. Furthermore, road traffic accidents affect the most productive members of society and result in large economic losses for a country. Therefore, it needs to create a way to predict the severity of the accident to manage the risk effectively. Additionally, the issue of deaths in Ethiopia is very serious, and it is crucial to identify the cause of the accident to mitigate traffic congestion. Machine learning (ML) is one technique to minimize this risk by analyzing historical data.

Machine learning is a subfield of computer science that enables computers to learn without explicit programming [11]. ML algorithms are executed on a given dataset to produce a model, representing the current dominant paradigm for machine learning. Researchers utilize ML to address accident severity prevention. In [12], an artificial neural network (ANN) was employed to explore the risk factors for injury severity in fatal crashes. The study involves binary classification of accidents with six attributes, achieving an accuracy of 77.5%. Reference [13] conducted modeling of road traffic accidents using a random forest algorithm, employing 15 attributes for model development with multiclass classification and achieving a score of 73.38%. Reference [14] also focused on accident severity prediction using a decision tree. The author utilized 15 attributes, and the algorithm yielded an accuracy of 73%. In [15], work zone crash injury severity was addressed using support vector machine (SVM), achieving an accuracy of 71.7% with six attributes in a multiclass classification task.

It is possible to note a gap in improving the model's performance using parameter tuning and identifying the best algorithm through data analysis. The aim of this study is to analyze road accident data to predict the severity of accidents. This assists police officers in understanding the causes of accidents, enabling informed decisions about transportation system rules and regulations. It also reduces the likelihood of accidents and saves lives. The contributions of the study are as follows:

- Developing a prediction model to predict accident severity using a new dataset
- Enhancing the model's performance using parameter tuning

The rest of the chapter is organized as follows: Sect. 2 describes the methods used, Sect. 3 presents the results, Sect. 4 discusses the findings, and Sect. 5 provides the conclusion, followed by references.

# 2 Materials and Methods

## *2.1 Dataset*

The sole data source for this study was the Department of Traffic Control and Investigation of the Addis Ababa Police Commission. With respect to this specific research, the author utilized data on road traffic accident records, which are stored in the traffic department. We collected a total of 7850 records with 18 attributes from the police department. The initial list of attributes from the police commission included roll number, collision, accident severity, date, time, sub-city, day, place, name of plaintiff, license plate numbers, name of defendant, sex, age, educational level, driving experience, investigator name and number of driving license, and decision or penalty. The dataset collected from the police commission has noisy and missing values. Therefore, the preprocessing techniques are crucial before proceeding to the analysis part.

## *2.2 Preprocessing*

Some attributes are not important for predicting the severity of the accident. Therefore, the authors removed the irrelevant attributes after discussion with domain experts in the field. Data cleaning techniques, including filling in missing values, removing null records, and eliminating outliers, were applied to address missing and noisy data. Additionally, attribute selection, data transformation, scaling, and labeling of the data were employed to prepare it for the next phase. After preprocessing, 7744 records and 6 attributes are used for the study. The final list of the attributes that were used for the study is shown in Table 1.

The statistical indicator of the strength of the relationship between two variables whose data correlation is presented in Fig. 1 is shown. It is also used to measure the extent to which two variables are linearly related.

The type of data for the attributes obtained from the traffic department is a mixture of strings and numbers. To analyze the data, it must first be labeled and transformed to make it usable by the selected tools and algorithm. Therefore, we labeled the data as shown in Table 2.

The proportionality of accident severity in the study is shown in Fig. 2.

**Table 1** List of attributes used in the study

| No. | Attribute name | Description |
|---|---|---|
| 1 | Accident place | The place where the accident occurred |
| 2 | Sex | The gender of the driver |
| 3 | Age | Age of the driver |
| 4 | Education level | The educational level of the driver |
| 5 | Driving experience | Driving experience of the driver |
| 6 | Accident type | The type of accident that occurred |

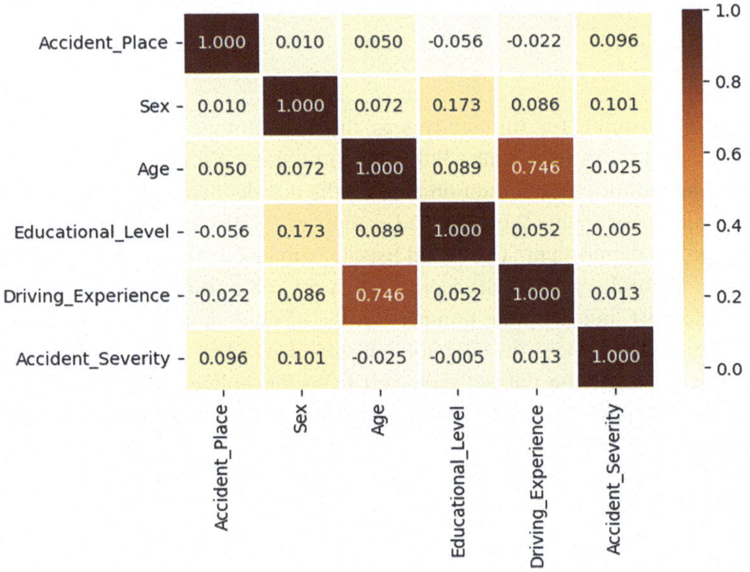


**Fig. 1** Correlation of the attribute

**Table 2** Description of the data

| No. | Attribute | Labeling description |
|---|---|---|
| 1 | Accident place | 22 = 1, 24 = 2, atlas = 3, ayat = 4, bole = 5, gerji = 6, goro = 7, gurdshola = 8, urael = 9, megenagna = 10, semit = 11, bsrate gebriel = 12, figa = 13, nfas silk = 14 |
| 2 | Sex | Male = 1, female = 2 |
| 3 | Age | 20–30 = 1, 31–40 = 2, 41–50 = 3, >50 = 4 |
| 4 | Educational level | 4–8 = 1, 9–12 = 2, diploma = 3, degree = 4, MSc and above = 5 |
| 5 | Driving experience | <1year=1, 1–5years = 2, 6–10 years = 3, 11–15 years = 4, 16–20 years = 5, 21–30 years = 6, and >30 years = 7 |
| 6 | Accident severity | Property damage = 1, light injury = 2, hard injury = 3, death = 4 |

To visualize the distribution of individuals in a different category, we use a bar graph depicted in Fig. 3.

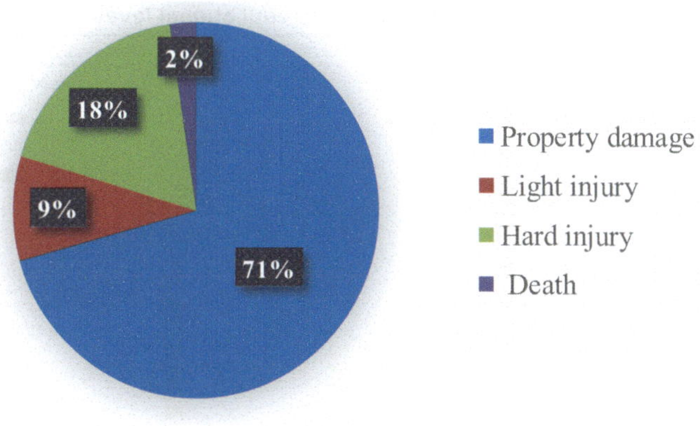


**Fig. 2** Accident severity proportionalities in the study

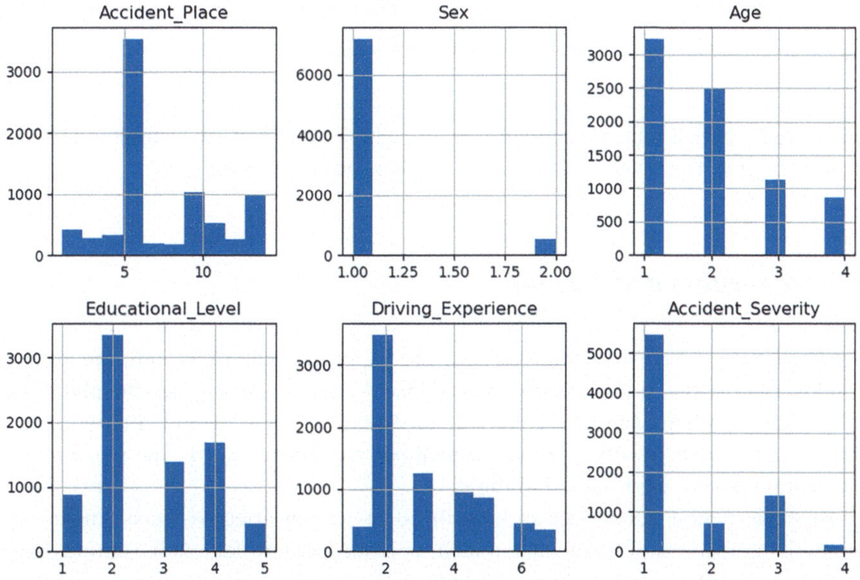


**Fig. 3** Distribution of individuals in a category

In Fig. 3, the x-axis is the attribute value like the place of the accident, sex, age, educational level, driving experience, and severity of the accident, while the y-axis is the value of each category in the data. The proportionality of the severity of the accident with the level of driver education is shown in Fig. 4.

Here, 1 represents the education level of the driver as 4–8, 2 corresponds to 9–12, 3 stands for diploma, 4 indicates a degree, and 5 is for MSc and above. In Fig. 4, most accidents occur when the driver's education level is marked with "2," which is between grades 9–12 followed by the label "4," which is the educational level of

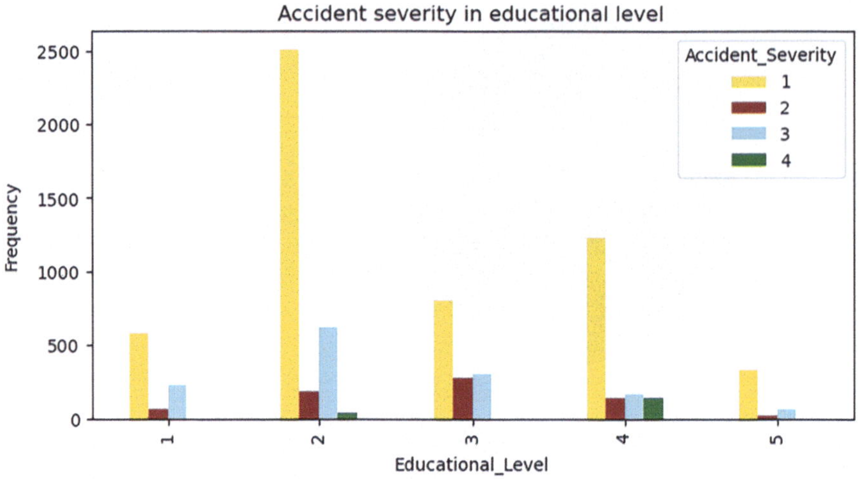


**Fig. 4** Proportionality of accident severity with educational level

the degree. There are fewer accidents in "5," where the driver's educational level is MSc or higher. The death also occurred at this educational level.

## 2.3 *Algorithm and Method*

*Artificial Neural Network.* With the aim of this experimental research, the author employed an artificial neural network (ANN). A neural network has the ability to infer complex relationships from ambiguous data, which is used to identify complex patterns and trends for other computer methods and humans [16]. The structure of the neural network is presented in Fig. 5.

Artificial neural networks can be utilized in several effective ways, including memorizing features of given data, match, or associate new data with existing data, learning rules or operations from examples, analyze and generalize from samples, and generating predictions. The ANN approach exhibits a good degree of tolerance for errors that may occur in the training set and is based on prior experience [17]. ANN is suitable for prediction due to hidden layers, allowing the model to capture nonlinear effect between relationships [18]. Each road accident severity attribute can be estimated using the sum product of the input-hidden and hidden-output connection weights. ANN has three layers: input layer, hidden layer, and output layer. Here, the input layer is related to the independent variables that have an impact on the severity of the accident. The hidden layer provides an interface between the input and output layers. However, the output layer is tied to the hidden layer according to the typical type of accident severity in the study. The feedforward-backward propagation neural network, which is widely used for prediction-type models and was thus used in this study, is one of the most popular ANN models.

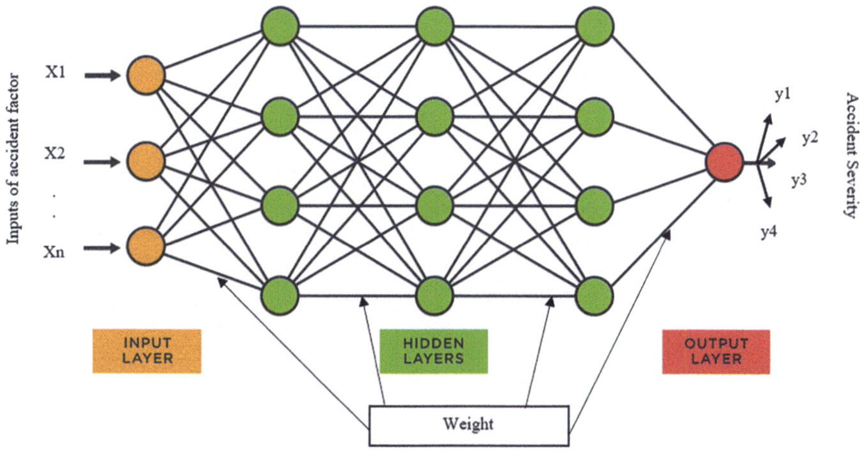


**Fig. 5** Structure of the artificial neural network

*Random Forest.* A supervised ensemble learning technique, called a random forest, operates according to the decision tree. It is extremely versatile and fast and can be used for classification tasks [19]. The random forest model utilizes an ensemble tree-based and learning algorithm, comprising a collection of classifier [20]. This ensemble approach involves averaging the predictions from various individual trees [21]. Instead of using the original sample, individual trees are constructed using bootstrap samples. It handles large datasets more effectively and reduces overfits [22]. RF is also robust in handling noisy data and imbalanced dataset. This is why we use RF for accident severity prediction.

*Decision Tree.* The decision tree is one of the common algorithms for classification tasks. It is a tree-based technique in which any path that starts from the root is described by a data separation sequence until a Boolean result is achieved at the leaf node [23]. It is a hierarchical illustration of the connections and nodes in knowledge relationships as internal decision nodes and terminal leaves [24]. The nodes represent objectives when relations are used to categorize. Based on the values of the input variables, the decision trees divide the data into a finite number of classes and work best with categorical data [25]. Instances and features are easily separated using the decision tree. We can quickly make a prediction using a decision tree, find an estimate of the results, and decide for future planning. Attributes are divided into subnodes as decision nodes in the decision tree [26]. The reason behind using DT is that they are good for handling mixed values in the data and are easy to understand.

*Support Vector Machine.* The support vector machine (SVM) is one of the most popular kernel-based learning algorithms that was first developed by Vapnik and his team in the late 1970s and is used in a variety of machine learning applications [27]. SVM is generally divided into two categories: support vector regression (SVR) and support vector classification. In many aspects of nonstationary data analysis, such

**Fig. 6** Flow of the study

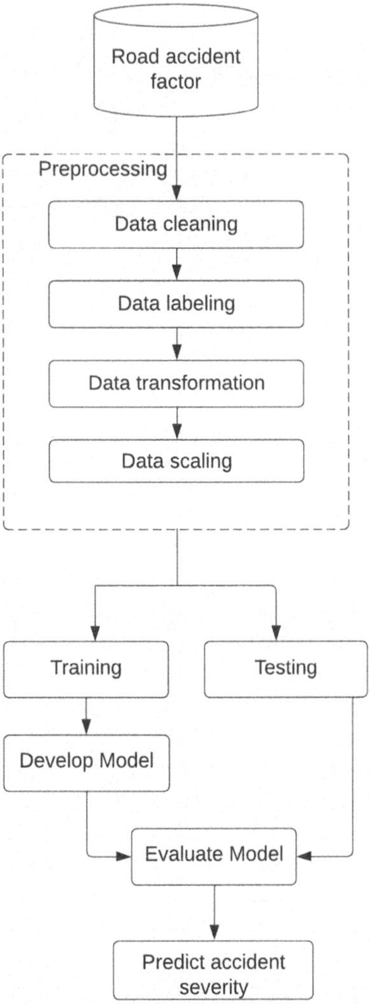


as restoring the cloaked forcing profile, estimating the relevance of input, modeling, and prediction, the work demonstrates improved precision [28]. SVM can linearly classify data samples into two groups. If the data cannot be separated linearly, the classification is carried out by SVM, which maps the data to a high-dimensional feature space [29].

*Tool.* In this study, the Python programming language and the Keras deep learning framework with the Jupyter notebook are used. The flow of the study is depicted in Fig. 6.

## 2.4 Evaluation Metrics

After developing the model, the next task is to evaluate the performance of the result. For this purpose, the author uses the confusion matrix, accuracy, precision, recall, F1 score, and Matthews correlation coefficient (MCC) to measure the performance of the model presented in Eqs. 1, 2, 3, and 4. [30].

| | | **Predicted** | |
|---|---|---|---|
| | | Positive | Negative |
| Actual | Positive | True positive (TP) | False negative (FN) |
| | Negative | False positive (FP) | True negative (TN) |

$$\text{Accuracy} = \frac{\text{TP} + \text{TN}}{\text{TP} + \text{FP} + \text{FN} + \text{TN}} \tag{1}$$

$$\text{Precision} = \frac{\text{TP}}{\text{TP} + \text{FP}} \tag{2}$$

$$\text{Recall} = \frac{\text{TP}}{\text{TP} + \text{FN}} \tag{3}$$

$$F1 - \text{Score} = 2^* \frac{\left(\text{Precision}^*\text{Recall}\right)}{(\text{Precision} + \text{Recall})} \tag{4}$$

$$\text{MCC} = \frac{(\text{TP} * \text{TN} - \text{FP} * \text{FN})}{\surd\,(\text{TP} + \text{FP})\,(\text{TP} + \text{FN})\,(\text{TN} + \text{FP})\,(\text{TN} + \text{FN})} \tag{5}$$

# 3 Results

For this study, the author uses 7744 records and 6 attributes, including the final class, to develop the model. The data were divided into two parts to train and test the model: 80% training and 20% testing, respectively. To do this, a machine learning technique with an algorithm of ANN, RF, DT, and SVM was used. To verify the performance of the model, a confusion matrix, accuracy, precision, and recall are used.

Keras is a powerful Python library for developing and evaluating deep learning models. In the ANN model, "sequential" as a model, "relu" as activation functions for the input and hidden layers, "softmax" as the output layer, "adam" as an

```
Model: "sequential"
_________________________________________________________________
 Layer (type)                Output Shape              Param #
=================================================================
 dense (Dense)               (None, 64)                384

 dense_1 (Dense)             (None, 64)                4160

 dense_2 (Dense)             (None, 56)                3640

 dense_3 (Dense)             (None, 36)                2052

 dense_4 (Dense)             (None, 4)                 148

=================================================================
Total params: 10,384
Trainable params: 10,384
Non-trainable params: 0
_________________________________________________________________
```

**Fig. 7** Model summary of ANN

**Table 3** The result of the study

| No. | Algorithm | Accuracy (%) | Precision (%) | Recall (%) | F1 score (%) |
|---|---|---|---|---|---|
| 1 | ANN | 90 | 90 | 90 | 90 |
| 2 | RF | 89 | 89 | 89 | 89 |
| 3 | DT | 87 | 87 | 87 | 86 |
| 4 | SVM | 89 | 89 | 89 | 88 |

**Table 4** Result of the study after using SMOTE analysis to balance the data

| No. | Algorithm | Accuracy | Precision | Recall | F1 score | MCC |
|---|---|---|---|---|---|---|
| 1 | ANN | 86 | 86 | 86 | 85 | 0.744 |
| 2 | RF | 86 | 90 | 86 | 87 | 0.746 |
| 3 | DT | 85 | 89 | 85 | 86 | 0.688 |
| 4 | SVM | 86 | 90 | 86 | 87 | 0.746 |

optimizer, "categorical_crossentropy" as a loss function, and "accuracy" as metrics were used in the study. The summary of the neural network model is shown in Fig. 7.

The result of the study before balancing the data is depicted in Table 3.

As can be seen in Table 3, the ANN algorithm performs slightly better than the other machine learning algorithms. There is a class imbalance in the severity of the accident. Therefore, to minimize this imbalance issue, we use SMOTE analysis to balance the data between the classes. The result of the algorithm after using the SMOTE analysis is depicted in Table 4.

As can be seen in Table 4, the result after the SMOTE analysis is almost similar in all machine learning algorithms. However, in RF and SVM the precision is better than the others. The confusion matrix of the result is depicted in Fig. 8.

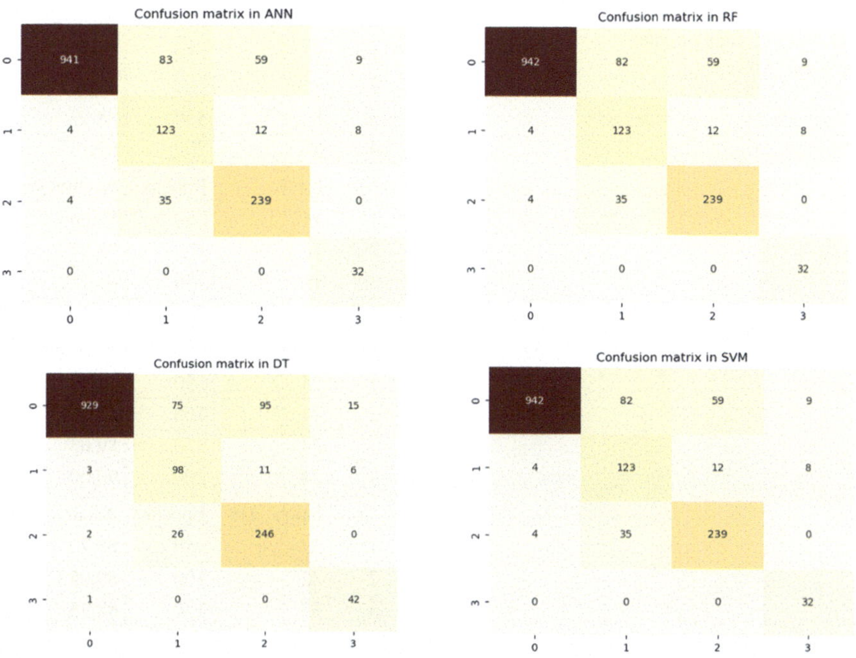


**Fig. 8** Confusion matrix of the study

In Fig. 8, the confusion matrix of the algorithms in ANN, RF, DT, and SVM after SMOTE analysis is presented. Of the 1549 traffic accident records, 1335, 1336, 1315, and 1336 traffic records are correctly classified in their predefined accident severity class, while 214, 213, 234, and 213 records are incorrectly classified in ANN, RF, DT, and SVM, respectively.

## 4 Discussion

After presenting the results, it is now time to discuss the findings. In our pursuit of improving the model's performance, we conducted parameter tuning; however, the best results are reported in this chapter. We assessed the model's performance by altering the training and testing data split, considering both 70% to 30% and 80% to 20% ratios. The latter, with an 80% training and 20% testing split, proved to be superior. For random forest, we found that setting the number of estimators to 70 and the max depth value 20 yielded the best result. In the case of decision tree, a max_depth of 20 and min_samples_split of 2 performed well. When using SVM, the kernel set to "rbf" and a value of c at 10,000 provided optimal results. Lastly, with ANN, we experimented with epoch values of 50, 100, 150, and 200, as well as batch_size values of 8, 16, 24, and 32 to identify the best configuration.

**Table 5** Result of the algorithm by changing the parameter value

| No. | batch_size | Epochs | Accuracy |
|---|---|---|---|
| 1 | 8 | 50 | 88.13 |
| 2 | 8 | 100 | 88.97 |
| 3 | 8 | 150 | 88.97 |
| 4 | 8 | 200 | 89.13 |
| No. | batch_size | Epochs | Accuracy |
| 1 | 16 | 50 | 88.60 |
| 2 | 16 | 100 | 88.87 |
| 3 | 16 | 150 | 88.71 |
| 4 | 16 | 200 | 89.15 |
| No. | batch_size | Epochs | Accuracy |
| 1 | 24 | 50 | 88.07 |
| 2 | 24 | 100 | 89.05 |
| 3 | 24 | 150 | 89.03 |
| 4 | 24 | 200 | 90.00 |
| No. | batch_size | Epochs | Accuracy |
| 1 | 32 | 50 | 88.20 |
| 2 | 32 | 100 | 89.05 |
| 3 | 32 | 150 | 88.95 |
| 4 | 32 | 200 | 89.12 |

**Table 6** Result of the algorithm with attribute scaling

| Algorithm | Accuracy | Precision | Recall | F1 score |
|---|---|---|---|---|
| ANN (scaling) | 90% | 79% | 89% | 83% |
| ANN (without scaling) | 89% | 76% | 89% | 81% |

Additionally, we explored different dropout rates in various layers, including 0.2, 0.4, and 0.5. However, the best outcome was achieved by setting the dropout value to 0.2 and applying it to the input and output layers. The result of these parameter variations is summarized in Table 5.

From Table 5, it is possible to observe that in most of the results, increasing the number of epochs is used to increase the performance of the model. Therefore, the batch_size of 24 and the epoch value of 200 scored well compared to the others. The result of the scaling algorithm and the attribute without scaling is shown in Table 6.

From the above output, scaling the attribute is important and had a good performance of 90% for accuracy, 79% for precision, 89% for recall, and 83% for the F1 score. ANN is used to predict the severity of the accident after splitting the dataset into training and testing sets at 80% and 20%, respectively. Attribute scaling was performed before parameter tuning of the result. As a result, machine learning algorithms are good at predicting the severity of accidents, and ANN is slightly better than the other algorithms. Figure 9 displays the accuracy and model loss of the study training and tests.

Table 7 shows the comparison of the previous work.

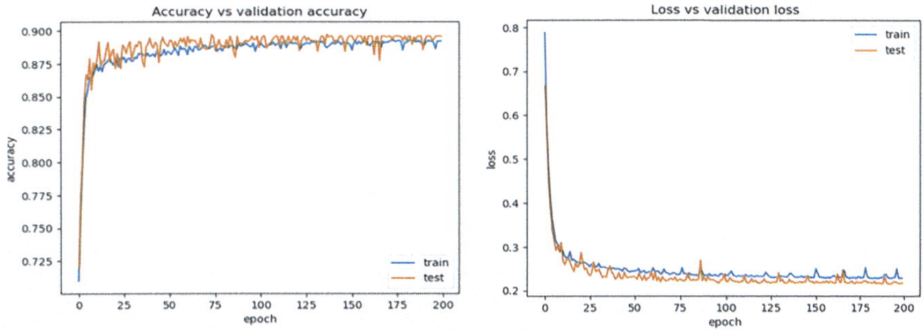


**Fig. 9** Accuracy and loss of the training and testing model

As can be seen in Table 7, ANN, RF, DT, and SVM in our study score better than in the previous. In [31], the overall prediction performance using ANN was 81.6% and 74.6% for the training and testing data. In [13], the author used 10 important selected attributes to obtain a better result score with an accuracy of 73.8 in the RF algorithm. Furthermore, the author in [14] classifies the accident into two classes with an accuracy of 73% in the DT algorithm. Finally, [15] use the kernel "rbf" and the 90% training and 10% testing test option using the SVM algorithm. The author got 71.7% accuracy and 80.4% precision. Therefore, using ANN, RF, DT, and SVM was a promising result in predicting accident severity with parameter tuning.

## 5 Conclusions

Road traffic accidents in Addis Ababa are a serious problem, and predicting their magnitude using consistent approaches has become a necessity. In this study, the type of accident severity that would provide life-saving safety has been predicted. Due to the nature of the data obtained from the traffic department, different data processing, labeling, and scaling were performed. For the study, 7744 records and 6 attributes were used to develop the model using the ANN, RF, DT, and SVM algorithms. The model result was 90% for accuracy, precision, recall, and the F1 score of the study. The ANN approach is used to model traffic accident data because it is a more flexible and assumption-free methodology and can evaluate and compare all aspects of traffic accidents. In addition to this scaling of the attribute, the performance of the result improved. As a result, ANN is a good algorithm for predicting the type of accident severity to reduce its occurrence. We did not find data related to lighting and road surface conditions. However, it is important to predict the severity of the accident. For future work, the author will include these variables and other machine learning algorithms to check the performance of the algorithms.

**Table 7** Comparison of our study with previous work

| Ref | Study place | No. of data | Class | Attribute | Algorithm | Acc | Pre | Rec | F1 |
|---|---|---|---|---|---|---|---|---|---|
| [31] | Abdhabi | 5973 | Multi | 16 | ANN | 74.6 | – | – | – |
| [12] | Kingdom SA | 12566 | Binary | 6 | ANN | 77.5 | – | – | – |
| [32] | Korea occupational | 6374 | Multi | 20 | RF | 71.3 | | | |
| [13] | Taiwan Highway | 9472 | Multi | 15 | RF | 73.38 | – | – | – |
| [14] | Bangldesh | 43089 | | 15 | DT | 73 | 71 | – | 70 |
| [33] | Wyoming Department of Transport | 1360 | Multi | 6 | DT | – | 79 | 74 | – |
| [15] | Florida Signal Four | 6865 | Multi | 6 | SVM | 71.7 | 80.4 | – | – |
| Our study | Addis Ababa Department of Transports | 7744 | Multi | 6 | ANN | 90 | 90 | 90 | 90 |
| | | | | | RF | 89 | 89 | 89 | 89 |
| | | | | | DT | 87 | 87 | 87 | 86 |
| | | | | | SVM | 89 | 89 | 89 | 88 |

## References

1. Traffic Collision. [Online]. Available: https://www.walshmedicalmedia.com/scholarly/traffic-collision-journals-articles-ppts-list-3345.html
2. Rajkumar, A.R., Prabhakar, S., Priyadharsini, A.M.: Prediction of road accident severity using machine learning algorithm. Int. J. Adv. Sci. Technol. **29**(6), 116–120 (2020)
3. Bazargani, H.S., Vahidi, R.G., Abhari, A.A.: Predictors of survival in motor vehicle accidents among motorcyclists, bicyclists and pedestrians. Trauma Mon. **22**(2), 4–10 (2017). https://doi.org/10.5812/traumamon.26019
4. Hayidso, T.H., Gemeda, D.O., Abraham, A.M.: Identifying road traffic accidents hots. Transp. Telecommun. [Online]. Available: https://www.longdom.org/scholarly/traffic-collision-journals-articles-ppts-list-3345.html
5. Road traffic injuries. [Online]. Available: https://www.who.int/news-room/fact-sheets/detail/road-traffic-injuries
6. Konlan, K.D., et al.: Prevalence and pattern of road traffic accidents among commercial motorcyclists in the Central Tongu District, Ghana. Sci. World J. **2020** (2020). https://doi.org/10.1155/2020/9493718
7. Woyessa, A.H., Heyi, W.D., Ture, N.H., Moti, B.K.: Patterns of road traffic accident, nature of related injuries, and post-crash outcome determinants in western Ethiopia – a hospital based study. African J. Emerg. Med. **11**(1), 123–131 (2021). https://doi.org/10.1016/j.afjem.2020.09.008
8. Fentahun, A.: Mining road traffic accident data for predicting accident severity to improve public health – role of driver and road factors in the case of Addis Ababa. MSc Thesis (2011)
9. Hanna, C.L., Hasselberg, M., Laflamme, L., Möller, J.: Road traffic crash circumstances and consequences among young unlicensed drivers: a Swedish cohort study on socioeconomic disparities. BMC Public Health. **10** (2010). https://doi.org/10.1186/1471-2458-10-14
10. World Health Organization: World Health Rankings. Accessed: 12 Nov 2022. [Online]. Available: https://www.worldlifeexpectancy.com/ethiopia-road-traffic-accidents
11. Bi, Q., Goodman, K.E., Kaminsky, J., Lessler, J.: What is machine learning? A primer for the epidemiologist. Am. J. Epidemiol. **188**(12), 2222–2239 (2019). https://doi.org/10.1093/aje/kwz189
12. Jamal, A., Umer, W.: Exploring the injury severity risk factors in fatal crashes with neural network. Int. J. Environ. Res. Public Health. **17**(20), 1–22 (2020). https://doi.org/10.3390/ijerph17207466
13. Chen, M.M., Chen, M.C.: Modeling road accident severity with comparisons of logistic regression, decision tree and random forest. Information. **11**(5) (2020). https://doi.org/10.3390/INFO11050270
14. Labib, M.F., Rifat, A.S., Hossain, M.M., Das, A.K., Nawrine, F.: Road accident analysis and prediction of accident severity by using machine learning in Bangladesh. In: 2019 7th Intetnational Conference Smart Computing and Communications. ICSCC 2019, pp. 7–11 (2019). https://doi.org/10.1109/ICSCC.2019.8843640
15. Mokhtarimousavi, S., Anderson, J.C., Azizinamini, A., Hadi, M.: Improved support vector machine models for work zone crash injury severity prediction and analysis. Transp. Res. Rec. **2673**(11), 680–692 (2019). https://doi.org/10.1177/0361198119845899
16. Çodur, A.T.M.Y.: An artificial neural network model for highway accident prediction: a case study of Erzurum, Turkey. Promet-Traffic Transp. **27**(3), 217–225 (2015). https://doi.org/10.12700/aph.11.05.2014.05.11
17. Muralitharan, K., Sakthivel, R., Vishnuvarthan, R.: Neural network based optimization approach for energy demand prediction in smart grid. Neurocomputing. **273**, 199–208 (2018). https://doi.org/10.1016/j.neucom.2017.08.017
18. Hussain, H.I., Kamarudin, F., Thaker, H.M.T., Salem, M.A.: Artificial neural network to model managerial timing decision: non-linear evidence of deviation from target leverage. Int. J. Comput. Intell. Syst. **12**(2), 1282–1294 (2019). https://doi.org/10.2991/ijcis.d.191101.002

19. Ghosh, M., et al.: A comparative analysis of machine learning algorithms to predict liver disease. Intell. Autom. Soft Comput. **30**(3), 917–928 (2021). https://doi.org/10.32604/iasc.2021.017989
20. Wubineh, B.Z., Metekiya, B.Y.: A knowledge-based system to predict crime from criminal records in the case of Hossana Police Commission. In: International Conference on Information and Communication Technology for Development for Africa (ICT4DA), pp. 90–95. IEEE (2023). https://doi.org/10.1109/ict4da59526.2023.10302252
21. Schonlau, M., Zou, R.Y.: The random forest algorithm for statistical learning. Stata J. **20**(1), 3–29 (2020). https://doi.org/10.1177/1536867X20909688
22. Khan, M.A., Memon, S.A., Farooq, F., Javed, M.F., Aslam, F., Alyousef, R.: Compressive strength of fly-ash-based geopolymer concrete by gene expression programming and random forest. Adv. Civ. Eng. **2021** (2021). https://doi.org/10.1155/2021/6618407
23. Charbuty, B., Abdulazeez, A.: Classification based on decision tree algorithm for machine learning. J. Appl. Sci. Technol. Trends. **2**(01), 20–28 (2021). https://doi.org/10.38094/jastt20165
24. Kassa, S.M., Wubineh, B.Z.: Use of machine learning to predict California bearing ratio of soils. Adv. Civ. Eng. **2023** (2023). https://doi.org/10.1155/2023/8198648
25. Tangirala, S.: Evaluating the impact of GINI index and information gain on classification using decision tree classifier algorithm. Int. J. Adv. Comput. Sci. Appl. **11**(2), 612–619 (2020). https://doi.org/10.14569/ijacsa.2020.0110277
26. Yadav, D.C., Pal, S.: Prediction of thyroid disease using decision tree ensemble method. Human-Intelligent Syst. Integr. **2**(1–4), 89–95 (2020). https://doi.org/10.1007/s42454-020-00006-y
27. Sheykhmousa, M., Mahdianpari, M., Ghanbari, H., Mohammadimanesh, F., Ghamisi, P., Homayouni, S.: Support vector machine versus random forest for remote sensing image classification: a meta-analysis and systematic review. IEEE J. Sel. Top. Appl. Earth Obs. Remote Sens. **13**, 6308–6325 (2020). https://doi.org/10.1109/JSTARS.2020.3026724
28. Raj, J.S., Ananthi, J.V.: Recurrent neural networks and nonlinear prediction in support vector machines. J. Soft Comput. Paradig. **2019**(1), 33–40 (2019). https://doi.org/10.36548/jscp.2019.1.004
29. Devikanniga, D., Ramu, A., Haldorai, A.: Efficient diagnosis of liver disease using support vector machine optimized with crows search algorithm. EAI Endorsed Trans. Energy Web. **7**(29), 1–10 (2020). https://doi.org/10.4108/EAI.13-7-2018.164177
30. Yassin, S.S., Pooja: Road accident prediction and model interpretation using a hybrid K-means and random forest algorithm approach. SN Appl. Sci. **2**(9), 1–13 (2020). https://doi.org/10.1007/s42452-020-3125-1
31. Alkheder, S., Taamneh, M., Taamneh, S.: Severity prediction of traffic accident using an artificial neural network. J. Forecast. **36**(1), 100–108 (2017)
32. Kang, H.R., Kyungsu: Predicting types of occupational accidents at construction sites in Korea using random forest model. Saf. Sci. **120**, 226–236 (2019)
33. Rezapour, M., Mehrara Molan, A., Ksaibati, K.: Analyzing injury severity of motorcycle at-fault crashes using machine learning techniques, decision tree and logistic regression models. Int. J. Transp. Sci. Technol. **9**(2), 89–99 (2020). https://doi.org/10.1016/j.ijtst.2019.10.002

# Part IV
# Artificial Intelligence and Machine Learning Applications

# Detection of Misinformation Related to Pandemic Diseases Using Machine Learning

**Javaria Naeem, Ömer Melih Gül, Ismail Burak Parlak, Kostas Karpouzis, Seifedine Nimer Kadry, and Yücel Batu Salman**

## 1 Introduction

The term "infodemic" has been coined to emphasize the dangers posed by the proliferation of misinformation during the management of disease outbreaks [1]. While there's abundant data on cybermedia platforms, the content can have both harmful and beneficial psychological effects on individuals [2]. Misinformation about the pandemic is a serious threat to public health and international relations [3]. For example, false claims linking 5G towers to COVID-19 have led to dangerous actions. In [4], the authors created "COVIDSENTI," a dataset with 90,000 COVID-19-related tweets categorized as positive, negative, and neutral. Our analysis highlights the significant influence of negative sentiment on public opinion.

J. Naeem · Ö. M. Gül
Department of Computer Engineering, Bahçeşehir University, İstanbul, Türkiye
e-mail: javaria.naeem@bahcesehir.edu.tr; omermelih.gul@bau.edu.tr

I. B. Parlak
Department of Computer Engineering, Galatasaray University, İstanbul, Türkiye
e-mail: bparlak@gsu.edu.tr

K. Karpouzis (✉)
Panteion Uninversity of Social and Political Sciences, Kallithea, Greece
e-mail: kkarpou@panteion.gr

S. N. Kadry
Department of Applied Data Science, Noroff College, Kristiansand, Norway

Artificial Intelligence Research Center (AIRC), Ajman University, Ajman, UAE
e-mail: seifedine.kadry@noroff.no

Y. B. Salman
Department of Software Engineering, Bahcesehir University, İstanbul, Türkiye
e-mail: batu.salman@bau.edu.tr



Ö. M. Gül et al. (eds.), *7th EAI International Conference on Robotic Sensor Networks*, EAI/Springer Innovations in Communication and Computing,
https://doi.org/10.1007/978-3-031-64495-5_11

Melton et al. [5] analyzed text data from 13 COVID-19 vaccine-related Reddit communities, ranging from December 1, 2020, to May 15, 2021; using sentiment analysis and LDA topic modeling, they tracked monthly sentiment shifts and latent issues. In [6], two datasets containing Instagram comments related to COVID-19 were used. These datasets span different time periods during the pandemic, allowing to examine the evolution of users' sentiments and perspectives concerning the epidemic. Lekshmi and Anoop [7] introduced a sentiment analysis strategy to assess public reactions to YouTube videos, employing text mining techniques and machine learning. Finally, authors in [8] compiled a multilanguage dataset consisting of tweets in English, German, French, and Greek and used BERT to train a sentiment analysis architecture, assisted by localized dictionaries to help with identifying slang, irony, or Proverbs and adages.

In this paper, we propose a sentiment analysis approach using comments on social media (YouTube, Instagram, Twitter, and Facebook). We applied deep learning and machine learning algorithms to these datasets and, in the end, we utilized additional algorithms (decision tree, logistic regression, and SVM using stacking classifier ensemble methodology) to improve accuracy. This comprehensive approach offers valuable insights into the dynamics of sentiment during the pandemic on social media, contributing to the ongoing discourse on addressing the challenges posed by the infodemic.

The subsequent sections of the paper are structured as follows: Sect. 2 includes a literature review, while Sect. 3 describes problem definition and explains the datasets we used. Section 4 explains the methodology that is used in this paper. Sections 5 and 6 discuss the results and present our conclusions.

## 2 Related Work

Within the literature, numerous studies delve into the examination of sentiment in website media posts, employing a diverse array of word-based evaluation techniques. The paper [9] analyzes COVID-19-related tweets from the start of the year. It uses tweets from various accounts, refines the data, and employs two datasets. Sentiments are assigned to the tweets, with WordCloud visualizations highlighting social media's limited effectiveness during the pandemic. The study [10] introduces an advanced approach to Twitter sentiment analysis of COVID-19. It focuses on Twitter users expressing their sentiments and classifies tweets as positive, negative, or neutral using multiple feature sets and classifiers. Performance evaluation, using ML and DL classifiers, shows impressive results with accuracy, precision, recall, and F1-score ranging from 93.88 to 96.66% in various datasets (COVIDSentiA, COVIDSentiB, and COVIDSentiC). In this research [11], Twitter analysis is done. They used N-gram piles autoencoder for trait extraction, and then, these features are classified using a decision tree, SVM, random forest, and k-nearest neighbors. The study [12] proposes an accurate sentiment analysis (SA) approach for COVID-19-related fake news. They use a contextual model with term weighting, applying eight ML and four DL algorithms. Based on the results, they create an effective Python prediction model. The model's performance is evaluated on unclassified COVID-19 fake news articles, predicting sentiment for each fabricated piece of information.This thesis embarks on a comprehensive investigation into sentiment analysis across prominent social media which covers Instagram, Facebook, YouTube, and Twitter, with a primary focus on discerning positive and negative information of COVID-19. The paper leverages a multifaceted approach, tailoring the choice of machine learning and deep learning algorithms to suit the specific characteristics of each platform. For Instagram and Facebook, machine learning techniques such as decision tree, logistic regression, and SVM are harnessed to extract sentiments from user-generated content. This methodological choice is guided by the nature of data prevalent on these platforms and the need for interpretable results. In contrast, the dynamic and nuanced nature of discourse on Twitter and YouTube necessitates the application of deep learning algorithms. For further boosting the accuracy and robustness of sentiment analysis, a novel approach is introduced, involving the integration of these machine learning and deep learning algorithms through stacking ensemble learning. This ensemble technique harnesses the strengths of each individual model to produce a refined and more reliable sentiment analysis framework. Through this research endeavor, our aim is to not only contribute to the field of sentiment analysis but also to provide valuable insights into the evolving public sentiments regarding COVID-19 across diverse social media platforms. By aligning our methodology with the unique characteristics of each platform, we endeavor to unravel the multifaceted nature of sentiments expressed in the digital realm during these unprecedented times.

## 3 Problem Definition

This work aims to perform sentiment analysis on diverse social platforms regarding COVID-19. It uses traditional ML methods for Facebook and Instagram and deep learning for Twitter and YouTube due to their unique characteristics. To enhance accuracy, a novel approach, stacking ensemble learning, integrates machine learning and deep learning. The results can benefit public health organizations, policymakers, and researchers during the ongoing pandemic.

### *3.1 Dataset*

We used four datasets extracted from YouTube, Twitter, Instagram, and Facebook for analysis. The datasets were obtained from Kaggle. The COVID-19 dataset is a collection of comments posted by Facebook, Instagram, Twitter, and YouTube users in response to various posts related to the COVID-19 pandemic.

### 3.1.1 YouTube Dataset

The size of the dataset is 42,851. The dataset contains a query, which contains keywords used to identify and gather comments related to COVID-19, including terms like "coronavirus," "COVID-19," "pandemic" and "vaccine," URL, title, upload date, channel, views, likes, dislikes, comment count, comment text, comment author, and comment date.

### 3.1.2 Twitter Dataset

The size of the dataset is 289,414. Dataset contains username: the username or display name of the twitter user who posted the tweet, user location, user created, user followers, user friends, user favorites, user verified, date, text, hashtags, and source.

### 3.1.3 Instagram Dataset

The size of the dataset is 1327. Dataset contains user id, username, comment id, comment text, profile URL, avatar URL, and date.

### 3.1.4 Facebook Dataset

The size of the dataset is 7029. Dataset contains date, Facebook ID, like count, profile ID, profile name, profile picture, profile URL, and text.

# 4 Methodology

Amid the exponential rise in coronavirus cases, a frantic race is underway among researchers and medical experts to develop quick diagnostics to effectively manage the spread of this disease. Situation has become increasingly dire, leading to a surge in fear and anxiety among individuals who are left wondering about the uncertain future.

## *4.1 Deep Learning*

In this deep learning methodology, we build a neural network using TensorFlow and Keras, consisting of several key layers. The process starts with the input layer, which serves as the entry point for your data, specifying its shape. Next, convolutional layer (Conv1D) is employed with 64 filters and a kernel size of 3, applying the ReLU activation function for feature extraction. Subsequently, a max pooling layer (MaxPooling1D) with a pool size of 2 is used to down-sample features. Global max pooling layer (GlobalMaxPooling1D) further lowers the spatial dimensions to a singular value for each feature map, obtaining the most significant features. Following this, a dense layer with 128 neurons, and ReLU activation is added for high-level feature representation. Finally, the output layer, configured with appropriate number of neurons corresponding to your output classes, employs softmax activation function for generating final predictions.

### *4.2 Machine Learning*

Machine learning methodology is a structured approach to solving real-world problems using machine learning techniques. Machine learning methodology is an iterative process that ensures systematic development and deployment of machine learning solutions. In this research, we applied three classifiers named logistic regression, SVM, and decision tree.

### *4.3 Logistic Regression*

Logistic regression is a fundamental and broadly used statistical and machine learning technique for binary and multi-class sorting problems. In this paper, the logistic regression model is trained by utilizing the training dataset. Following the training phase, the model is employed to make predictions on the test dataset.

### *4.4 SVM (Support Vector Machine)*

Support vector machine (SVM) is a versatile machine learning algorithm applicable to tasks involving classification and regression. SVMs handle linear and nonlinear problems and are valuable in, text analysis. In this paper, SVM is used to classify the COVID-19 comments. The first dataset is divided into features and target labels and then split into testing and training data. The svc object is then trained by employing the training dataset with the fit method. This phase finds the optimal decision boundary that separates the data points into separate categories while optimizing the margin between them. After training the SVM classifier, predictions are generated for the test dataset.

### 4.5 Decision Tree

A decision tree organizes data in a hierarchical tree-like format. In this paper, a decision tree classifier is used to analyze the comment of COVID-19. First, feature matrix and target matrix are created from dataset. Employing the fit method, decision tree classifier is trained with the training data. During training, the decision tree algorithm constructs a tree structure that divides the data iteratively according to the most informative features, with the objective of minimizing impurity or maximizing information gain. After training, the decision tree classifier makes prediction on test data.

### 4.6 Proposed Approach

We used the stacking classifier ensemble methodology. We used logistic regression classifier as the meta-model. The primary function of the meta-model is to merge predictions from base models, resulting in a final prediction. Logistic regression is selected for its simplicity and interpretability, making it an appropriate choice for aggregating predictions. The stacking classifier is trained on the training data. Throughout the training process, each of the base models is fitted to the training data to generate individual predictions. When the training process is finished, the stacking classifier is tasked with predicting outcomes on the test dataset. The ensemble combines the predictions from the base models using the meta-model (logistic regression) to produce the final predictions. The accuracy of the stacking classifier's predictions is evaluated by comparing them to the true labels in the test dataset. The accuracy score is calculated using the accuracy_score function. Ensemble learning methodology is given in Fig. 1 (Fig. 2).

## 5 Numerical Results

We share the outcome of our analysis of the COVID-19 sentiment dataset gathered from diverse social networking sites like Instagram, Facebook, Twitter, and YouTube. Our analysis focused on identifying positive and negative sentiments related to COVID-19. We applied deep learning on Twitter and YouTube and machine learning on Facebook and Instagram.

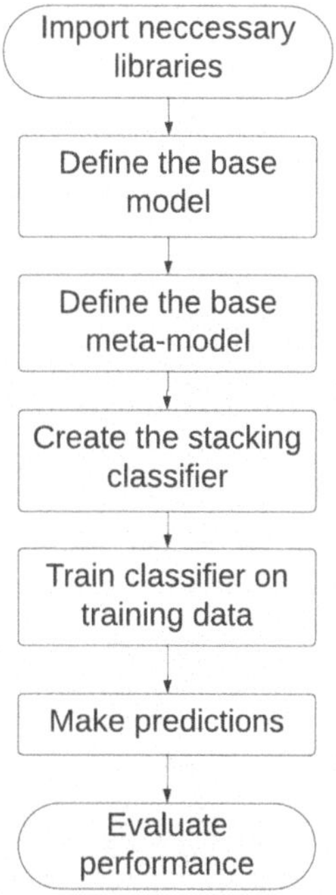


**Fig. 1** Ensemble learning methodology

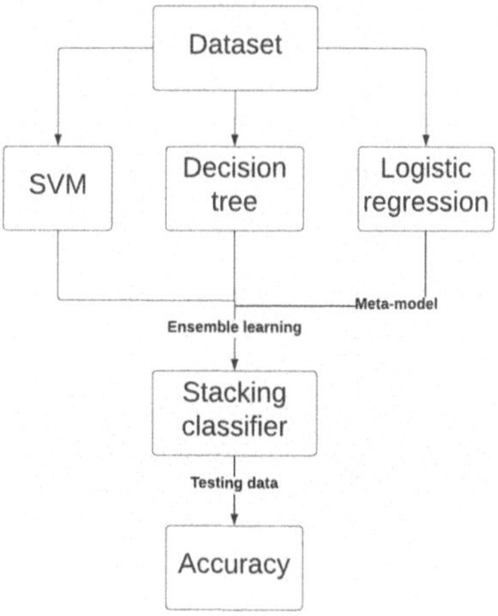


**Fig. 2** Ensemble learning methodology framework

## *5.1 Facebook*

The Facebook dataset contains 7029 comments. For analysis, three machine learning including decision tree, logistic regression, and SVM are applied to the Facebook dataset. Our aim is to evaluate the performance of these algorithms. To do so, datasets are partitioned into training and testing subsets. The training set is 90% and the testing set is 10% of the dataset.

### 5.1.1 Logistic Regression

By applying logistic regression, we get an accuracy of 84.7%. The classification report can be found in Table 1.

In the case of negative class, the precision is 0.88, indicating that model's predictions of negative are accurate 88% of the time. In the case of "neutral" class, the recall score is 0.96, indicating the model accurately detected 96% of the actual "neutral" instances. In the "positive" class F1-score is 0.88 signifying a favorable equilibrium between precision and recall. The model's overall accuracy stands at 85% indicating that it made correct classification for 85% of the total instances of the dataset.

### 5.1.2 SVM

By applying SVM, we get an accuracy of 76.9%, as shown in Table 2.

For the "negative" class, precision is 0.88. This implies that when model makes a prediction of "negative" for an instance, it is accurate roughly 88% of the time. For the "neutral" class, the precision is 0.75. For *"neutral"* predictions, the model is accurate in roughly 75% of cases. Regarding the "positive" class, a precision score of 0.93% demonstrates model's accuracy in predicting "positive" around 93% of the time. In "negative" class, a recall of 0.75 implies that model precisely pinpointed

**Table 1** Classification report of Facebook posts using logistic regression

| | Precision | Recall | F1 | Score |
|---|---|---|---|---|
| Negative | 0.88 | 0.75 | 0.81 | 167 |
| Neutral | 0.75 | 0.96 | 0.84 | 187 |
| Positive | 0.93 | 0.83 | 0.88 | 236 |
| Accuracy | 0.85 | 590 | | |
| Macro avg | 0.86 | 0.85 | 0.84 | 590 |
| Weighted avg | 0.86 | 0.85 | 0.85 | 590 |

**Table 2** Classification report of Facebook posts using SVM

| | Precision | Recall | F1 | Score |
|---|---|---|---|---|
| Negative | 0.53 | 0.89 | 0.66 | 99 |
| Neutral | 0.93 | 0.71 | 0.80 | 244 |
| Positive | 0.82 | 0.78 | 0.80 | 247 |
| Accuracy | 0.77 | 590 | | |
| Macro avg | 0.76 | 0.79 | 0.75 | 590 |
| Weighted avg | 0.81 | 0.77 | 0.78 | 590 |

**Table 3** Classification report of Facebook posts using decision trees

| | Precision | Recall | F1 | Score |
|---|---|---|---|---|
| Negative | 0.80 | 0.78 | 0.79 | 167 |
| Neutral | 0.87 | 0.98 | 0.93 | 187 |
| Positive | 0.88 | 0.81 | 0.85 | 236 |
| Accuracy | 0.86 | 590 | | |
| Macro avg | 0.85 | 0.86 | 0.85 | 590 |
| Weighted avg | 0.86 | 0.86 | 0.86 | 590 |

75% of actual "negative" instances. In "neutral" class, a recall score is 0.96 indicates that the mode accurately recognized 96% of actual "neutral" instances. For the "positive" class, recall is 0.83, meaning that the model correctly finds 83% of the actual "positive" instances. For the negative class, F1-score is 0.81. For neutral class, F1-score is 0.84. For positive class, F1-score is 0.88. The result of the model is 0.85; it correctly classified 85% of the total instances in the dataset.

#### 5.1.3 Decision Tree

By applying decision tree classifier, we get an accuracy of 76.9%. Table 3 shows the classification results.

Precision of 0.80 for the "negative" class indicates that the model anticipates "negative." It is precise in roughly 80% of instances. Model predicts "negative" for an instance in "negative" segment, it is accurate approximately 80% of time. For the "neutral" section, precision is 0.87, clarify that it is correct about 87%. Precision value for "positive" class is 0.88%. In case "negative" class, 0.78 is the recall score, indicating 78% of the actual "negative" instances. For the "neutral" class, recall is 0.98, suggesting 98% of the actual "neutral" instances. For the "positive" section, recall is 0.8. The F1-score for negative, neutral, and positive are 0.79, 0.93, and 0.85. With an overall accuracy of 0.86, the model correctly classified 86% of total instances in the dataset. After applying ensemble learning, achieved accuracy increases to 91%.

### *5.2 Instagram*

The Instagram dataset contains 1327 comments. Machine learning techniques including decision tree, logistic regression, and SVM are applied to Instagram dataset. Our aim to evaluate the performance of these algorithms. To do so, in the dataset, 90% is allocated for training, and 10% is designated for testing.

### 5.2.1 Logistic Regression

By applying logistic regression, we get an accuracy of 74.3%. The classification report is shown in Table 4.

For the "negative" class, the precision is 0.14. This shows that prediction for "negative" instance is correct about 14% of the time. The precision score for "neutral", and "positive" is 0.98, and 0.59 indicates that the efficiency of model finding positive is 98 and 59%. In the "negative" class, the model achieved a recall of 0.50, meaning it accurately identified half of the actual "negative" instances. When considering the "neutral" class, a recall score of 0.72 indicates that the model accurately recognized 72% of the "neutral" instances. The "positive" class, recall is 0.85 demonstrating model identifies 85% of "positive" occurrences. F1-score for "negative", "neutral," and "positive" class is 0.22, 0.83, and 0.70, respectively. The model's accuracy rate of 0.74 denotes that it made accurate classifications for 74% of total instances in data.

### 5.2.2 Decision Tree Classifier

By applying decision tree classifier, we get an accuracy of 76.9%. Table 5 shows the results for this algorithm.

The precision of "negative" class is 0.14; it suggests that the model is correct about 14% of the time when "negative." Regarding to "neutral" class, score of precision is 0.98 demonstrates model's accuracy in predicting 'neutral' about 98% while "positive" class, precision is 0.59, suggesting, prediction of "positive," case is 59%. Recall for "negative," "neutral," and "positive" class; recall is 0.44, 0.93, and 0.89. The F1-score for classes, "negative," "neutral," and "positive" is 0.46,

**Table 4** Classification report of Instagram posts using logistic regression

| | Precision | Recall | F1 | Score |
|---|---|---|---|---|
| Negative | 0.14 | 0.50 | 0.22 | 4 |
| Neutral | 0.98 | 0.72 | 0.83 | 82 |
| Positive | 0.59 | 0.85 | 0.70 | 27 |
| Accuracy | 0.74 | 113 | | |
| Macro avg | 0.57 | 0.69 | 0.58 | 113 |
| Weighted avg | 0.86 | 0.74 | 0.78 | 113 |

**Table 5** Classification report of Instagram posts using decision tree

| | Precision | Recall | F1 | Score |
|---|---|---|---|---|
| Negative | 0.50 | 0.44 | 0.46 | 16 |
| Neutral | 0.93 | 0.80 | 0.86 | 70 |
| Positive | 0.62 | 0.89 | 0.73 | 27 |
| Accuracy | 0.77 | 113 | | |
| Macro avg | 0.68 | 0.71 | 0.69 | 113 |
| Weighted avg | 0.80 | 0.77 | 0.77 | 113 |

0.86, and 0.73. There are four instances of "negative." The efficiency of the model is 0.74, showing that it correctly identified 74% of the total instances in the dataset.

### 5.2.3 SVM

By applying SVM, we get an accuracy of 69.9%, as shown in Table 6.

The precision result for "negative" is 0.50 which explains that model predicts an instance as "negative," is valid about 50%. Precision of "neutral" class is 0.93, shows when model is predicted as "neutral," 93%. "Positive" class attained a precision of 0.62 which means 62%. In the case of "negative" class, the recall is 0.44, signifying the model accurately detected 44% of negative cases. The recall score for "neutral" and "positive" is 0.80 and 0.89. It shows, model accurately identified 80% of 'neutral' occurrences, 89% of "positive" instances. the "negative" class. The F1-scores of "neutral" and "positive" are 0.78 and 0.68. The total accuracy of the framework is 0.77, which means that it correctly identified 77% of the total instances in the dataset. The accuracy obtained after ensemble learning is 79%.

## 5.3 *Twitter*

The Twitter dataset contains 289,414 tweets; after applying the deep learning network, we obtain the results shown in Table 7.

There are three classes tagged as 0, 1, and 2; the recall score for Class 0 is 0.74, while for Class 1 recall is 0.86. The third class (Class 2) gets a 0.78 score of recall,

**Table 6** Classification report of Instagram using SVM

| | Precision | Recall | F1 | Score |
|---|---|---|---|---|
| Negative | 0.00 | 0.00 | 0.00 | 0 |
| Neutral | 0.90 | 0.69 | 0.78 | 78 |
| Positive | 0.64 | 0.71 | 0.68 | 35 |
| Accuracy | 0.70 | 113 | | |
| Macro avg | 0.51 | 0.47 | 0.49 | 113 |
| Weighted avg | 0.82 | 0.70 | 0.75 | 113 |

**Table 7** Classification report of Twitter dataset

| | Precision | Recall | F1 | Score |
|---|---|---|---|---|
| Negative | 0.78 | 0.74 | 0.74 | 5406 |
| Neutral | 0.76 | 0.81 | 0.81 | 8703 |
| Positive | 0.85 | 0.81 | 0.81 | 8037 |
| Accuracy | | 0.79 | 0.79 | 22,146 |
| Macro avg | 0.78 | 0.78 | 0.78 | 22,146 |
| Weighted avg | 0.80 | 0.79 | 0.79 | 22,146 |

**Table 8** Classification report of YouTube comment dataset

| | Precision | Recall | F1 | Score |
|---|---|---|---|---|
| Negative | 0.91 | 0.89 | 0.90 | 1190 |
| Neutral | 0.96 | 0.96 | 0.96 | 1931 |
| Positive | 0.93 | 0.95 | 0.94 | 1997 |
| Accuracy | | | 0.94 | 5118 |
| Macro avg | 0.94 | 0.93 | 0.78 | 5118 |
| Weighted avg | 0.94 | 0.94 | 0.79 | 5118 |

meaning that the model correctly identified 78% of the tweets. Classes 0, 1, and 2 generate an F1 score equal to 0.74, 0.81, and 0.81, respectively. The final accuracy obtained by the model is 79%. After applying the ensemble learning, we get 80% accuracy.

### 5.4 *YouTube*

The YouTube dataset contains 42,851 comments. After removing empty or nonsensical entries, the deep learning network is applied (Table 8).

When the model identifies Class 0, accuracy is about 91%, while for Class 1, precision is 0.96, suggesting that Class 1 is correct for about 96% of the posts. Finally, for Class 2, precision is 0.93, meaning that the model is correct at about 93%. Class 0 achieved recall 0.89, illustrating that the model identified 89% of the actual instances. Class 1 and 2 gained recall 0.96 and 0.95 suggesting that the model correctly identified 96 and 95% of the factual occurrences. Class 0 achieved an F1-score of 0.90, while Class 1 attained 0.96 and Class 2 reached 0.94. The complete accuracy of the model is 0.94, revealing that it successfully identified 94% of the total instances in the dataset. After applying ensemble learning, accuracy is improved to 95%.

## 6 Conclusion

During the COVID-19 pandemic, this study analyzed sentiment on Instagram and Facebook using conventional machine learning methods and employed deep learning techniques for Twitter and YouTube due to their unstructured content. The research introduced stacking ensemble learning to enhance sentiment analysis accuracy by combining machine and deep learning models; this method proved to be the best method for improving the accuracy for Facebook, Twitter, Instagram, and YouTube content, improving detection accuracy to 91% on Facebook, 79% for Instagram, 95% of YouTube, and 80% for Twitter.

Future research can investigate the use of robotic agents or *chatbots* to automatically fact-check and verify information posted on social media; these robots could interact with users to provide real-time information verification. In the same context, research can investigate the role of social bots in spreading fake news within online social networks and analyze their impact on robotic communication systems, developing strategies to detect and mitigate their influence.

## References


1. Cinelli, M., Quattrociocchi, W., Galeazzi, A., Valensise, C.M., Brugnoli, E., Schmidt, A.L., Zola, P., Scala, A.: The COVID-19 social media infodemic. Sci. Rep. **10**(1), 1–10 (2020)
2. Alamoodi, A.H., Zaidan, B.B., Zaidan, A.A., Albahri, O.S., Mohammed, K.I., Malik, R.Q., Almahdi, E.M., Chyad, M.A., Tareq, Z., Albahri, A.S., Hameed, H., Alaa, M.: Sentiment analysis and its applications in fighting COVID-19 and infectious diseases: a systematic review. Expert Syst. Appl. **167**, 114155 (2021)
3. Roozenbeek, J., Schneider, C.R., Dryhurst, S., Kerr, J., Freeman, A.L., Recchia, G., van der Bles, A.M., Van Der Linden, S. (2020). Susceptibility to misinformation about COVID-19 around the world. R. Soc. Open Sci. **7**(10), 201199 (2021)
4. Naseem, U., Razzak, I., Khushi, M., Eklund, P.W., Kim, J.: COVIDSenti: a large-scale benchmark Twitter data set for COVID-19 sentiment analysis. IEEE Trans. Comput. Soc. Syst. **8**(4), 1003–1015 (2021)
5. Melton, C.A., Olusanya, O.A., Ammar, N., Shaban-Nejad, A.: Public sentiment analysis and topic modeling regarding COVID-19 vaccines on the Reddit social media platform: a call to action for strengthening vaccine confidence. J. Infect. Pub. Health **14**(10), 1505–1512 (2021)
6. Karayiğit, H., Akdagli, A., Acı, Ç.İ.: BERT-based transfer learning model for COVID-19 sentiment analysis on Turkish instagram comments. Inf. Technol. Control **51**(3), 409–428 (2022)
7. Lekshmi, S., Anoop, V.S.: Sentiment analysis on COVID-19 news videos using machine learning techniques. In: Proceedings of International Conference on Frontiers in Computing and Systems: COMSYS 2021, pp. 551–560 (2022)
8. Skarpelos, Y., Messini, S., Roinioti, E., Kaperonis, S., Marazoti, M.G., Karpouzis, K., Tzouveli, P.: Social and emotional responses to COVID-19: A comparative semiotic approach. In: XIII International Conference of Semiotics Semiotics Across and In-Between, 2023 (2023)
9. Chakraborty, K., Bhatia, S., Bhattacharyya, S., Platos, J., Bag, R., Hassanien, A.E.: Sentiment analysis of COVID-19 tweets by deep learning classifiers—a study to show how popularity is affecting accuracy in social media. Appl. Soft Comput. **97**, 106754 (2020)
10. Jalil, Z., Abbasi, A., Javed, A.R., Badruddin Khan, M., Abul Hasanat, M.H., Malik, K.M., Saudagar, A.K.J.: COVID-19 related sentiment analysis using state-of-the-art machine learning and deep learning techniques. Front. Pub. Health **9**, 812735 (2022)
11. Kandasamy, V., Trojovský, P., Machot, F.A., Kyamakya, K., Bacanin, N., Askar, S., Abouhawwash, M.: Sentimental analysis of COVID-19 related messages in social networks by involving an N-gram stacked autoencoder integrated in an ensemble learning scheme. Sensors **21**(22), 7582 (2021)
12. Bangyal, W.H., Qasim, R., Rehman, N.U., Ahmad, Z., Dar, H., Rukhsar, L., Aman, Z., Ahmad, J.: Detection of fake news text classification on COVID-19 using deep learning approaches. Comput. Math. Methods Med. **2021**, 1–14 (2021)

# Test the Capability of Arduino TinyML for Machine Learning

**Valentin Caravella, Sahar Yassine, and Seifedine Nimer Kadry**

## 1 Introduction

TinyML is nowadays considered a new approach to artificial intelligence usage. The goal is to use machine learning with low power, small memories, and low size, aiming toward the usage in many projects to revolutionize our conception of the world [1]. With TinyML, it is now possible to introduce machine learning into new uses for typical everyday objects. Nonetheless, we can also imagine and create new objects with this technology, which could simplify numerous problems. For example, we can imagine its use in an industrial context: Mechanical problems could be identified by sound recognition. The device recognizes normal and usual sounds typical of the industrial system, and when there is an unrecognizable sound, it will be able to signal the anomaly. With TinyML, we can imagine many very useful uses with very few resources. Using frugal machine learning may be an answer to the resource's restrictions in the near future. We could use it for tasks allowing us to get rid of any computer-driven help.

Many microcontrollers have been developed to allow most people to develop their own TinyML applications. Arduino is the famous board brand that developed

V. Caravella
Polytech Marseille, Marseille, France

S. Yassine (✉)
Noroff University College, Kristiansand, Norway
e-mail: Sahar.Yassine@noroff.no

S. N. Kadry
Department of Applied Data Science, Noroff College, Kristiansand, Norway

Artificial Intelligence Research Center (AIRC), Ajman University, Ajman, UAE

 
Ö. M. Gül et al. (eds.), *7th EAI International Conference on Robotic Sensor Networks*, EAI/Springer Innovations in Communication and Computing,
https://doi.org/10.1007/978-3-031-64495-5_12

Arduino Nano 33 BLE Sense board. This board is able to use a neural network, which is equipped with numerous sensors and takes advantage of the Arduino architecture to exploit other non-integrated sensors. With Arduino and Google, Harvard University, as part of their EdX online course, is associated with creating a very great Tiny Machine Learning Kit with Arduino Nano 33 BLE Sense board. This makes it easier to understand how to use TinyML.

In this chapter, the goal is to illustrate the capability of Arduino Nano 33 BLE Sense board for machine learning, and the tests will be supported by Harvard EdX courses and their examples. By showing the capability for machine learning of this board, it would be possible that more people will be likely to use it.

Subsequently, the capability of Arduino TinyML will be tested by comparison between Arduino program examples and computer program equivalents. The tests are color recognition, keyword spotting, and gesture recognition.

## 2 State of the Art

TinyML is a method that can be adopted for resource-limited settings by providing portability, intelligence, and a simple protocol that integrates with the preexisting IoT devices [2]. There are only a few papers available on the use of TinyML. In one of the research efforts [3], the researchers demonstrated a review of the possibilities of using TinyML, listed the tools that support TinyML, and presented the potential of improving TinyML systems.

Another study [4] identified the challenges and opportunities of developing truly smart frugal objects. The study discussed the available TinyML frameworks and presented a real case study that resulted by proposing a multi-radio access network (RAT) architecture for smart frugal objects.

However, very few talk specifically about the Arduino Nano 33 BLE Sense board. In the following sections, we will discuss TinyML existing benchmarks, a Tiny Machine Learning course provided by Harvard EdX, and TensorFlow Lite for embedded systems.

### *2.1 Benchmarking of TinyML Systems*

In a research attempt [5], thanks to the benchmarking of TinyML systems, we can observe what is possible using TinyML. Various inputs on board are possible, including audio, images, behavioral measurements, and industrial telemetries. Numerous applications are possible with only these examples, and this is just an illustration of the comparison to differentiate some of the TinyML boards. Furthermore, this document presents a comparison between processors and does not deal with specific TinyML boards, so it may be worth taking a closer look at the boards specifically.

Another research effort [6] designed three types of fully connected neural networks (NNs), trained each of them using 10 datasets, and demonstrated the benchmark by reporting the onboard model performance on seven popular MCU boards similar to the ones used to design TinyML hardware.

### *2.2 Tiny Machine Learning Course*

The paper [1] is associated with the Harvard EdX course. We can observe a fully presented course structure and their teaching methods for introducing machine learning with TinyML. Moreover, we can see a comparison between Pico4ML, Arduino Nano 33 BLE Sense, and STMicroelectronics Sensor Tile boards, and understand why the Arduino board is a very good choice to use for TinyML projects. The Arduino board has all the sensors needed for embedded systems, and the Arduino software is a great asset, with its numerous libraries and IDE, which are ideal for beginners. Furthermore, Harvard courses give access to numerous machine learning examples on a computer and a few on Arduino, perfect for understanding how it works. We will use their examples in our tests. This chapter, however, does not comment on the performance of Arduino results.

### *2.3 TensorFlow Lite for Embedded System*

The paper [7] deals with TensorFlow Lite Micro, a version of TensorFlow for microcontrollers. TensorFlow is a library for machine learning created by Google. On the one hand, TensorFlow Lite is used to transform a machine learning model into a lighter version using quantified data, in particular over a specific range $[-128;127]$. It aims to develop its use on mobile and embedded devices. On the other hand, TensorFlow Lite Micro is specifically focused on microcontrollers and embedded systems with limited resources. The aim is to learn a model with TensorFlow on a computer, and then deploy it in microcontrollers with TensorFlow Lite Micro, which is a lighter version of the model. TensorFlow Lite Micro allows the user to convert the model into hexadecimal format, and the microcontroller's version of TensorFlow Lite Micro allows us to interpret the hexadecimal model.

## 3 Materials and Methods

To carry out the tests of the capability of Arduino TinyML, we will use a computer equipped with Python and Arduino Nano 33 BLE Sense. It is necessary to use a computer to do machine learning with Arduino because we need to train the model on the computer that will then be implemented into Arduino. In Table 1, we may observe the characteristics between the Arduino board and the computer used during tests.

**Table 1** Characteristics of the equipment used

| System name | CPU | Clock | Memory | Price |
|---|---|---|---|---|
| Arduino Nano 33 BLE Sense | 32-bit ARM Cortex-M4F | 64 MHz | 1 MB flash256 kB RAM | 22.80 € |
| Computer | Intel Core i7-8700K | 3.70 GHz | 500 Go SSD32 Go RAM | >799.00 € |

Arduino characteristics: https://store.arduino.cc/products/arduino-nano-33-ble

The computer characteristics are those of the host laboratory/the computer price is a starting price based on the laptop used a few times.

The software used for the tests are Google Collab and Arduino IDE version 1.8.19. With Google Collab, it is easy to use with TensorFlow, and in addition, we can use the data to analyze on Google Drive. For the Arduino IDE, it was more relevant to work with the old IDE of Arduino because its Serial Monitor is located in another window, unlike the new IDE.

To perfectly use Arduino Nano 33 BLE Sense board, we need to install several libraries and the board manager. The Arduino Mbed OS Nano Boards allow our board to be used in the Arduino IDE as it is not supported by default in this environment. For libraries, we needed Arduino_TensorFlowLite, but this version was deprecated by Google. Harvard has integrated this library into their Harvard_TinyMLx library with the example we used for our tests. To use all sensors, we need to install Arduino_LSM9DS1 library for accelerometer, magnetometer, and gyroscope sensors, ArduinoBLE library for the Bluetooth module, and Arduino_APDS9960 library for the color sensor.

Finally, to test as many things as possible on the board, the programs and datasets set up are based on existing programs and data, which we will mention later. Arduino is based on the use of integrated sensors in contrast to computer systems, whose primary use is not similar. It is necessary to find a compromise to illustrate Arduino's advantage in real time with what is usually possible with a computer. This is why some tests are different between the Arduino and the computer because the Arduino tests are based on sensor-based data acquisition, while the computer tests are based on the use of classic datasets. Please note that all models, whatever the nature of the data, must be trained by a computer, only then can the data be adapted and finally integrated into the code to be uploaded to Arduino. The data to test model are always tested in the same order, following a precise test plan to compare computer and Arduino results. For more details about the Arduino program and the used datasets, please check https://github.com/SaharYassine/TinyML.

### *3.1 Color Recognition*

Color recognition is a test whose aim is to make a comparison between Visual Wake Words, and it represents a common TinyML visual use case of identifying whether an element is present in the image or not [1] on the computer and color classification

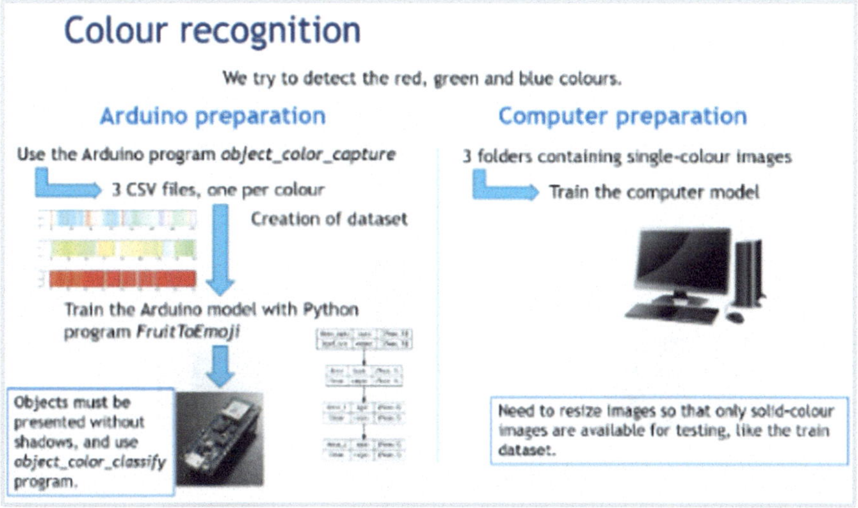


**Fig. 1** Illustration of how to use the Arduino board and a computer

by color sensor on Arduino (Fig. 1). We will try to make the system recognize the correct primary color among the three primary colors (blue, red, and green). For the computer version, the program used is based on the Visual Wake Words assignment from the Harvard EdX course [1], and the model is trained and validated by two unicolored datasets. For the Arduino version, the dataset is collected by Arduino itself, the output is printed in an Arduino serial monitor, and we need to copy and paste it into a document in .txt format.

We need to do that for each color. These data are trained by a neural network on a computer, then reduced in size and extracted in hexadecimal format for integration into Arduino. My test will consist of collecting data and testing it into the models. My test data are based on five objects by color. My experiment is based on me showing them to the Arduino color sensor and taking photos of them to use with the computer model. Please also note that the Arduino sensor detects in RGB format and shadows can be a source of error, which is why the object and Arduino board must be presented vertically under a light source every time for optimal result of detection. Also note that for the computer version we need to resize the images so that only full-color images are available for testing, such as the training dataset. The expected results for the two versions are to have two color detection spectrums for each color tested, so we use them to make confusion matrices and a comparison between them to illustrate the successful recognition of the colors for each version.

## 3.2 *Keyword Spotting*

Keyword spotting refers to the word's identification in an audio format. Our test is aimed at making a relevant comparison between audio recognition of words by a computer and Arduino (Fig. 2).

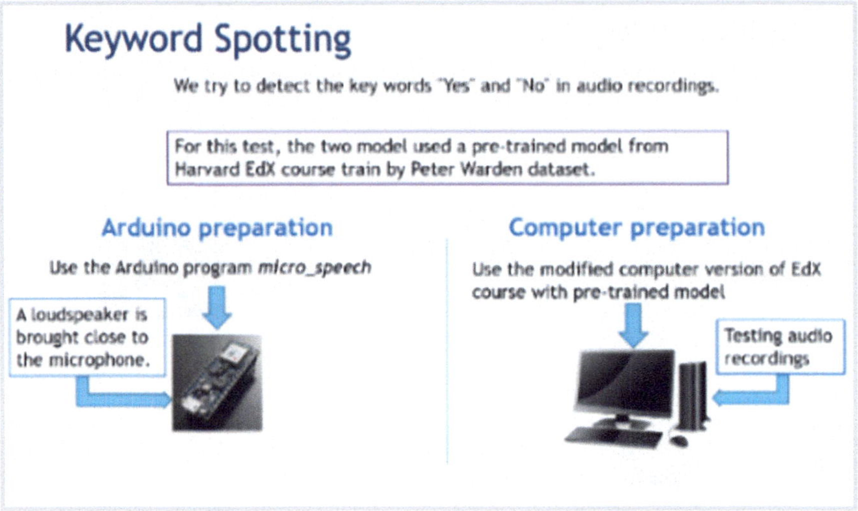


**Fig. 2** Illustration of how to use the Arduino board and a computer

For this part, we face version compatibility problems with TensorFlow library with 1.x version and its 2.x version, which is why we use a pretrained model from the Harvard EdX course [1]. This pretrained model is the same for both computer and Arduino, and is also a quantified model, ready to be adapted for the Arduino board. The Arduino program used is a micro speech example from the Harvard_TinyMLx library. The model has been trained to recognize "Yes" and "No" words with the Pete Warden dataset for word recognition [8]. The output of this model is "Silence," "Unknown," "Yes," and "No." To check whether the model recognized the real "Yes" and "No," a few volunteers created and sent us audio recordings of unknown words of their choice to ensure it was correctly categorized as "Unknown." Data are collected from these volunteers using the .wav format to be ready to test on a computer. To use these audios on Arduino, a speaker is approached close to the Arduino microphone. The expected results for this test are to get at least two word detection spectrums amongst all audios tested, and finally a comparison between them to illustrate the successful recognition of correct words.

### *3.3 Gesture Recognition*

The gesture recognition is based on the magic wand program from the Arduino example of Harvard_TinyMLx library. The concept of this program is to draw a number in the air using both the accelerometer and gyroscope. Then the Arduino board draws the movement into two dimensions. Afterward, this drawing is tested in a neural network to guess what number it may be (Fig. 3).

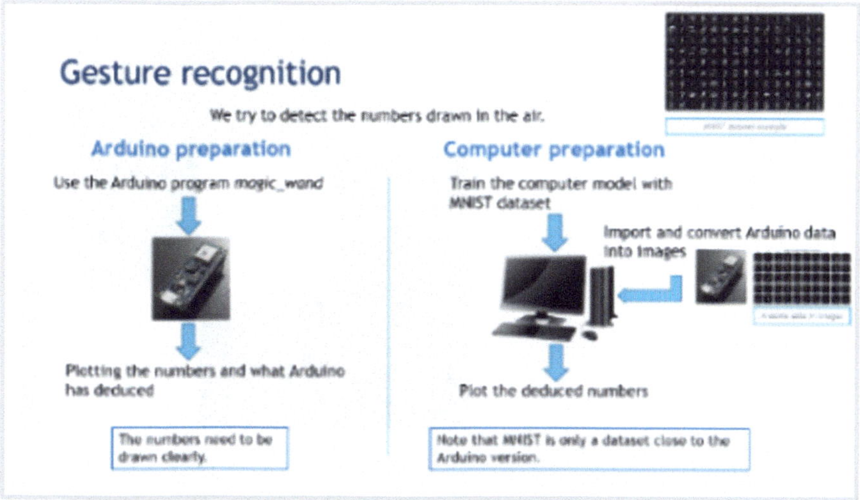


**Fig. 3** Illustration of how to use the Arduino board and a computer

The aim of this test is to show that, even without an equivalent program on a computer, we can still deal with the nature of the Arduino output. Arduino writes in the serial monitor the drawing traced and the number it guessed from it appears on the screen. If the operator is applied correctly, we can get close to the number in handwriting. There is actually a dedicated dataset for recognizing handwritten numbers: MNIST dataset. Therefore, for this test we can try to use the MNIST dataset to train a computer model and analyze number recognition with it. The outputs in the serial monitor are alphanumeric characters. The first step is to create images with a black background like the MNIST dataset. It is known that Arduino would most likely give better results than computers because we need to have a clear image of the drawing in output to be sure that the computer can detect numbers like the MNIST dataset. After processing into an image, and if the numbers are recognizable enough, it will be likely that we get a clear image. Most of the time, the Arduino board recognized it. Finally, the results for this test are: two number detection spectrums for all numbers drawn in the air tested and a comparison between them to see which number has the best results on both Arduino and a computer.

## 4 Results and Discussion

After running our various tests, we obtained the following results.

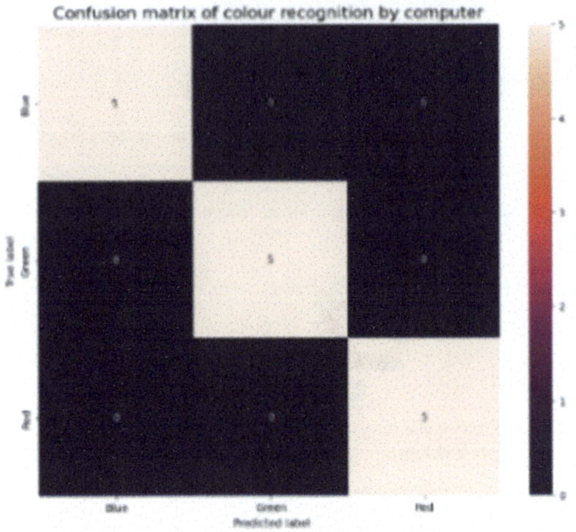


**Fig. 4** Confusion matrix of color recognition by a computer

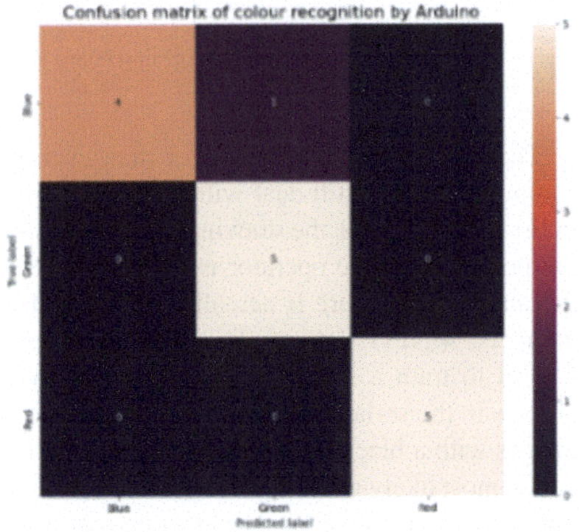


**Fig. 5** Confusion matrix of color recognition by Arduino

## *4.1 Color Recognition Results*

In our color recognition test, our results are very satisfactory. Note that we must be in good condition to ensure the correct reading of colors without shadows for the Arduino version and resizing photos to obtain a single-color image for a computer.

The results of color recognition by a computer with CNN are shown in Fig. 4. The computer manages to recognize all the images perfectly this time. During other attempts, the computer fails for a blue object. In Fig. 5, these are the results of the Arduino version with a dense-layer regression network based on the FruitToEmoji project. This time, Arduino succeeds almost perfectly, but fails for a blue object with a false-negative green; it is the same object as mentioned for the computer version in the previous tests (Fig. 6).

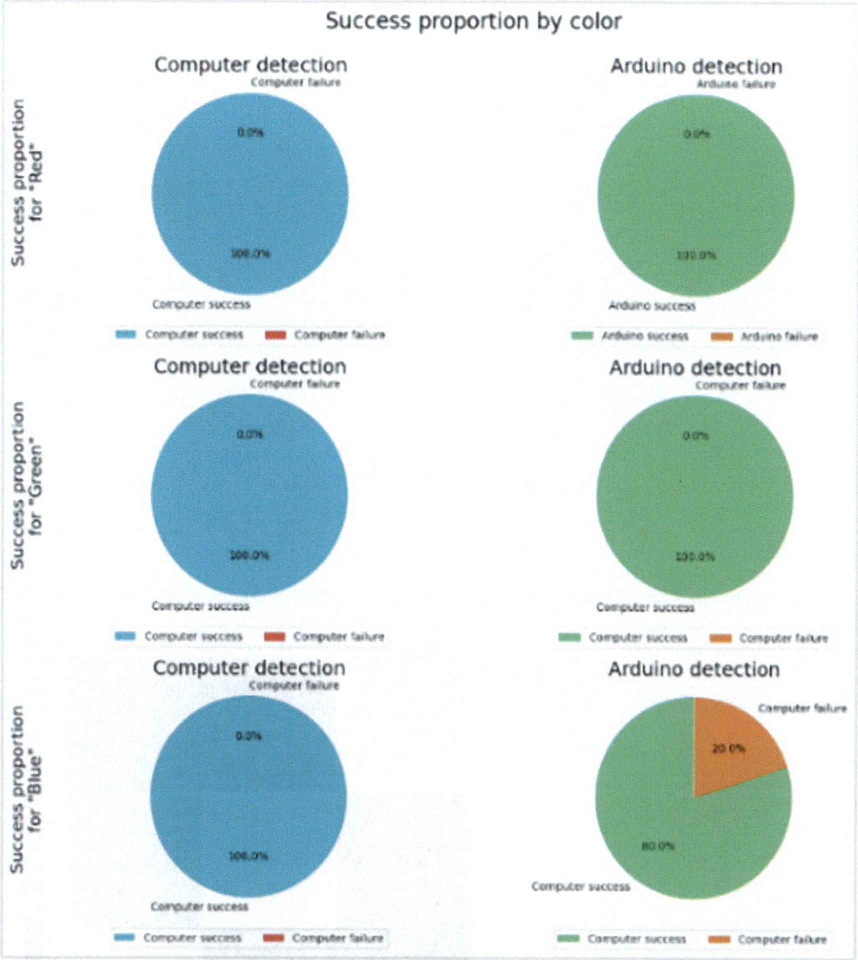


**Fig. 6** Pie charts of success proportion for each color recognition with a computer and Arduino board

Figure 7 illustrates the results of a computer and Arduino board about color recognition with a computer and Arduino. Finally, the computer has a 100% success rate compared with 93.3% for the Arduino board, based on our test data.

## 4.2 *Keyword Spotting Results*

In keyword spotting tests, the model is the same for the computer and Arduino versions. The model is pretrained by the Harvard EdX course to recognize "Yes"

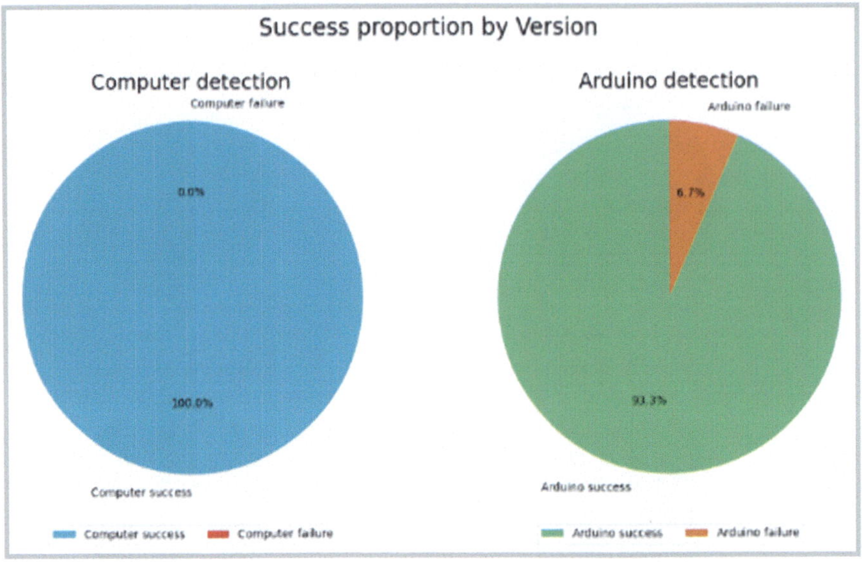


**Fig. 7** Pie charts of success proportion for color recognition with a computer and Arduino board

**Fig. 8** Confusion matrix of audio recognition by a computer

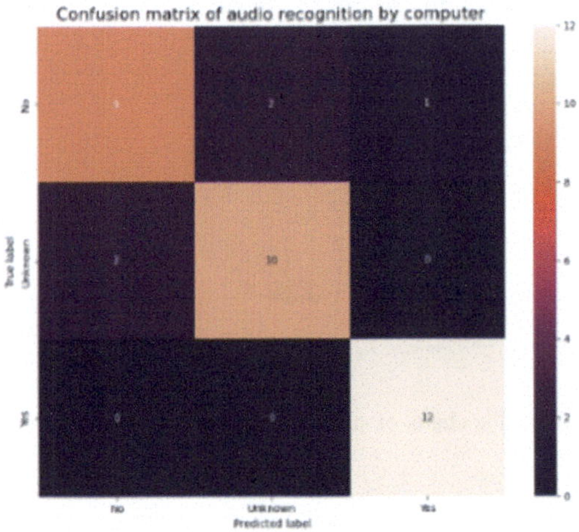


and "No,"; this is why it is interesting to see when the model makes mistakes with an unknown word.

Figure 8 shows the computer's recognition, and we can see that only the "Yes" word is perfectly recognized. Some "No" and "Unknown" words were not recognized. "No" is recognized as a false negative "Unknown" and "Yes," and "Unknown" words are recognized as a false negative for "No."

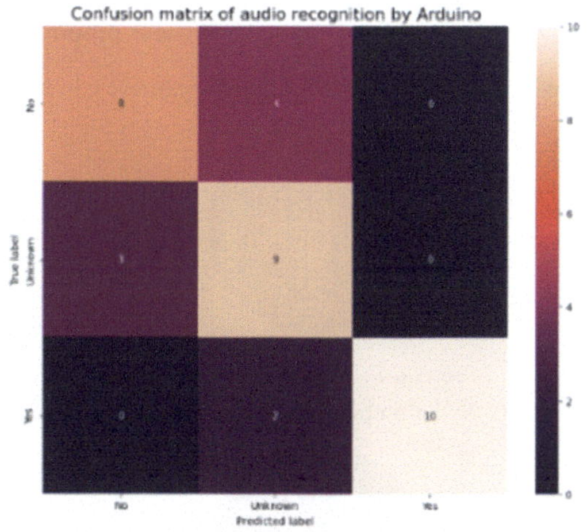


**Fig. 9** Confusion matrix of audio recognition by Arduino

The Arduino's recognition is shown in Fig. 9. This time there are more false negatives but only one type for each word class. The "No" words are the class with the most recognition error (Fig. 10).

Finally, according to the previous results and Fig. 11, the computer allows better recognition of words by audio, with an overall success rate of 86.1% against 75.0% for the Arduino's detection. Nevertheless, we can see that the error is not excessively high compared to that of the computer.

## *4.3 Gesture Recognition Results*

As a reminder, in gesture recognition the two models are different but, in both cases, the input to the models is an image with a handwritten number. In the computer case, the model is trained using the MNIST dataset; and in the Arduino case, the model uses a self-made dataset of Arduino data.

In Fig. 15, we can see the Arduino's data for the tests converted into black background images to be read on a computer with model training from the MNIST dataset. Figure 12 shows the results of this data in the computer model; some numbers are recognizable, but the majority are not, or even number classes are never recognized and only make false negatives.

Figure 13 shows Arduino's number recognition, and we can see that most numbers are recognized, but like the computer model, a few numbers produce false negatives. Only number 3 is perfectly recognized.

If we now look at the success rate for each model (Fig. 14), we can see that the computer makes a lot of mistakes with a success rate of 48%, which can be explained by the original nature of the training data and test data. Arduino's success rate is 72%. Finally, this test shows that it is possible to use Arduino to make gesture

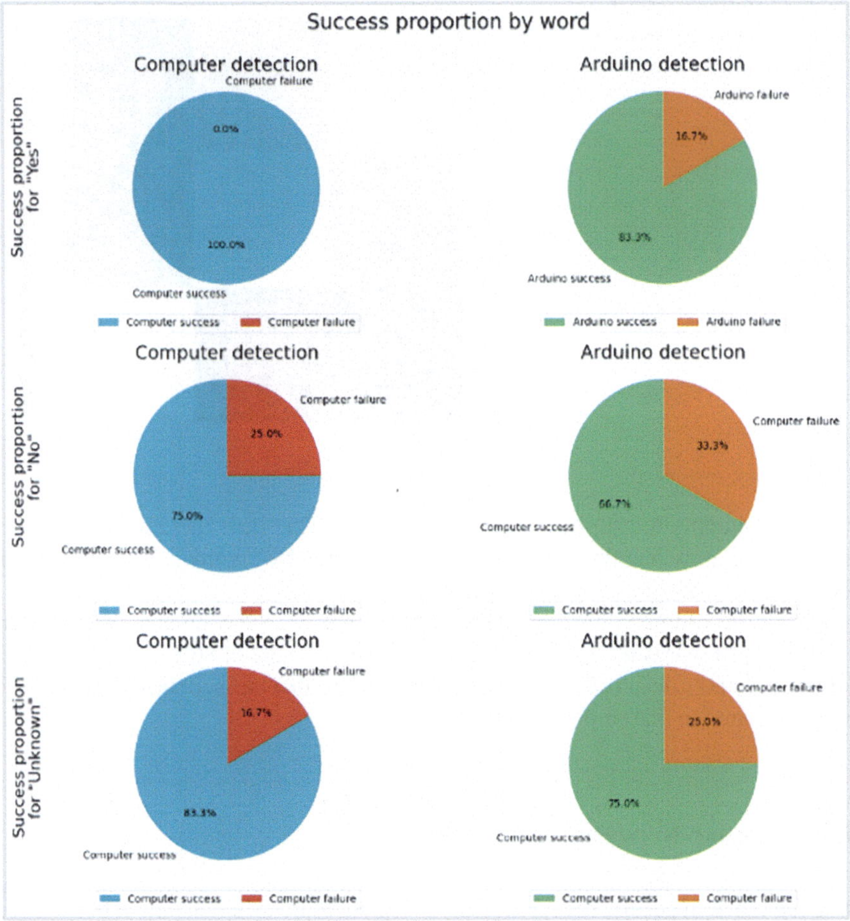


**Fig. 10** Pie charts of success proportion for each word in the speech recognition with a computer and Arduino board

recognition illustrated by the recognition of numbers in this case, so the error is important but with the nature of this data, it is a similar problem for the computer model (Fig. 15).

## 5 Conclusions

In conclusion, we see that the Arduino Nano 33 BLE Sense board generally produces worse results than a computer. However, the results are not excessively different, and we can imagine that if the Arduino model is trained with good data, the

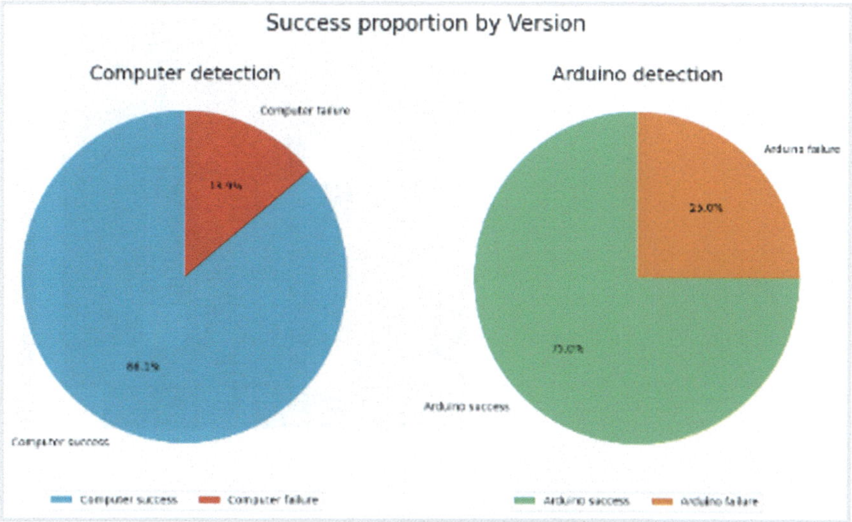


**Fig. 11** Pie charts of success proportion for audio recognition with a computer and Arduino board

**Fig. 12** Confusion matrix of number recognition by computer

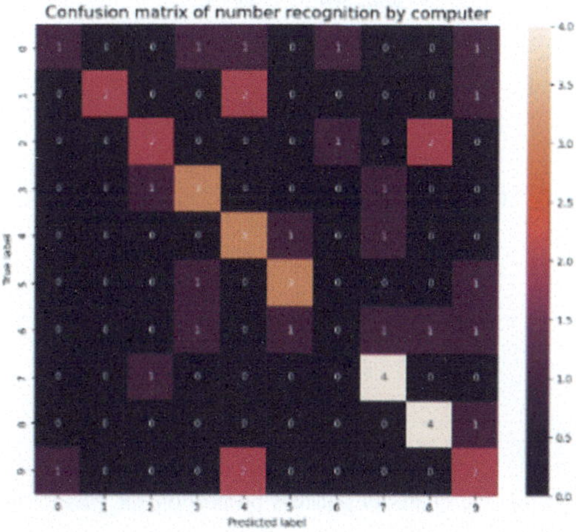


Arduino board will be relevant. Finally, by using the TinyML concept, this Arduino board makes it possible to create machine learning applications at a low cost and with a few resources. In a context where AI is being introduced more and more into our daily lives and research, and where fossil resources are running out, it is interesting to have this possibility of an alternative.

However, to use this Arduino board, it is necessary to resize the applications. The capacities are limited, and the use cases for this type of device fall within the scope of intelligent sensors that could be used in Industry 4.0. Also, with this Arduino

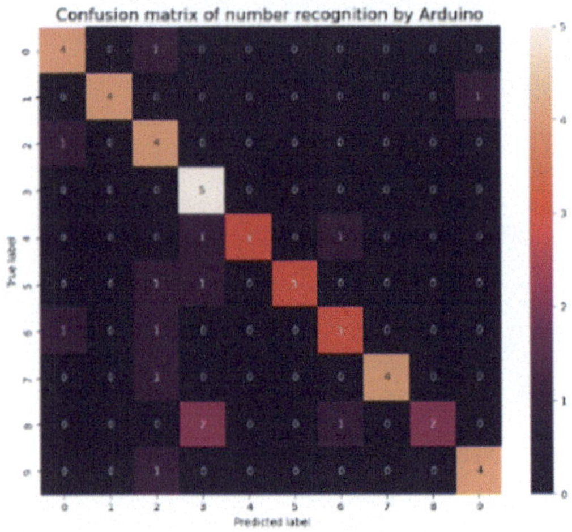


**Fig. 13** Confusion matrix of number recognition by Arduino

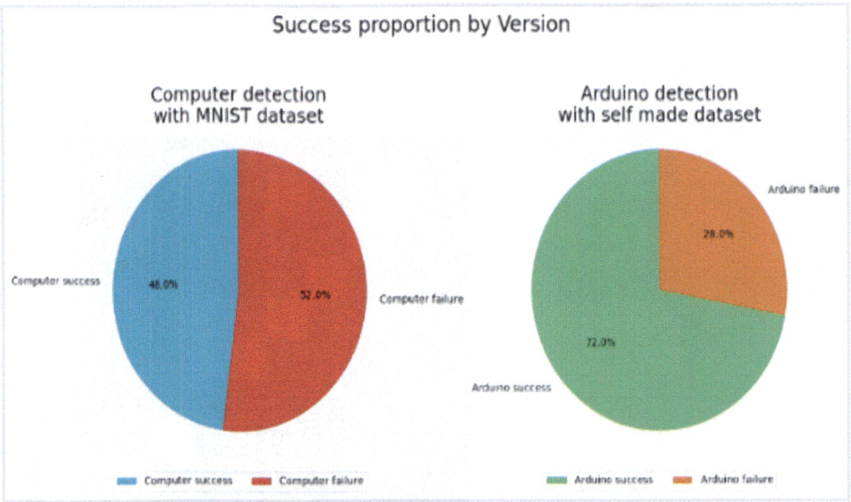


**Fig. 14** Pie charts of success proportion for gesture recognition with a computer and Arduino board

board, we cannot use it like a computer and cannot import external data into the board.

To improve this research, we need to become more specialized in developing neural networks on Arduino. The examples used are complex to be modified, and the many changes in the Python and Arduino libraries make it difficult to train new models without first specializing in this area.

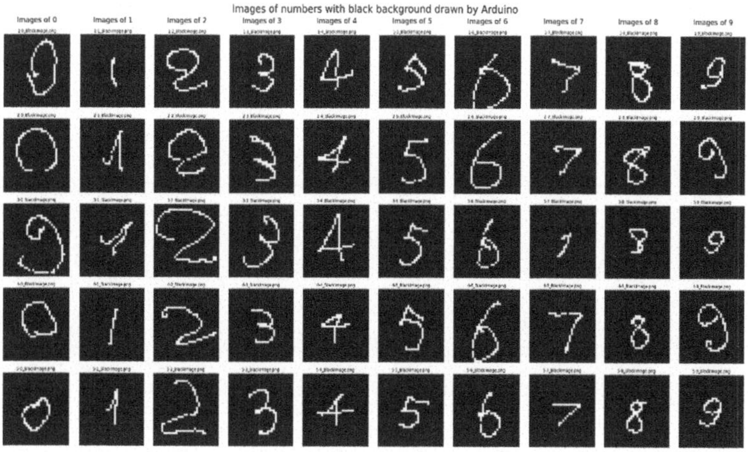


**Fig. 15** Image of Arduino data converted to MNIST data format

# References

1. Reddi, V.J., et al.: Widening Access to Applied Machine Learning with TinyML [Online]. Available: http://arxiv.org/abs/2106.04008 (2021)
2. Dutta, D.L., Bharali, S.: TinyML meets IoT: a comprehensive survey. Internet of Things (Netherlands). **16**. Elsevier B.V. (2021). https://doi.org/10.1016/j.iot.2021.100461
3. Ray, P.P.: A review on TinyML: State-of-the-art and prospects. J. King Saud University – Comput. Inf. Sci. **34**(4) King Saud bin Abdulaziz University, 1595–1623 (2022). https://doi.org/10.1016/j.jksuci.2021.11.019
4. Sanchez-Iborra, R., Skarmeta, A.F.: TinyML-enabled frugal smart objects: challenges and opportunities. IEEE Circuits Syst Mag. **20**(3), 4–18 (2020). https://doi.org/10.1109/MCAS.2020.3005467
5. Banbury, C. R., et al.: Benchmarking TinyML Systems: Challenges and Direction. Accessed: 8 Sept 2023. [Online]. Available: http://arxiv.org/abs/2003.04821 (2020)
6. Sudharsan, B., et al.: TinyML benchmark: executing fully connected neural networks on commodity microcontrollers. In: 7th IEEE World Forum on Internet of Things, WF-IoT 2021, pp. 883–884. Institute of Electrical and Electronics Engineers Inc. (2021). https://doi.org/10.1109/WF-IoT51360.2021.9595024
7. David, R. et al.: TensorFlow Lite Micro: Embedded Machine Learning on TinyML Systems [Online]. Available: http://arxiv.org/abs/2010.08678 (2020)
8. Warden, P.: Speech Commands: A Dataset for Limited-Vocabulary Speech Recognition [Online]. Available: http://arxiv.org/abs/1804.03209 (2018)

# Comparison of Different Methods for Estimation of Arterial Blood Pressure Using PPG Signals

**Teodora Mladenovska, Nevena Ackovska, Magdalena Kostoska, Bojana Koteska, Katarina Trojachanec Dineva, and Ana Madevska Bogdanova**

## 1 Introduction

An essential indicator that can provide vital information about an individual's health is blood pressure (BP). Significantly affecting cardiovascular health, blood pressure is quantified as the force that blood exerts against the vessel walls. Blood pressure is determined by two values: systolic pressure (SBP) and diastolic pressure (DBP) [1]. The SBP and DBP, respectively, can be derived from the maximal and minimal values of an ABP signal. One method for measuring Arterial Blood Pressure (ABP) involves an invasive approach. Aimed to perform a real-time blood pressure monitoring, this technique requires peripheral artery cannulation and is usually associated with some complications such as bleeding or infection, since it is very sensitive to body movement. Cuff-based methodologies have been widely utilized for measuring arterial blood pressure (ABP). However, the use of these measures may lead to patient discomfort during the phases of cuff inflation and deflation. Also, this process has the ability to impact the degree of accuracy and create additional layers of uncertainty. As a result, researchers and healthcare practitioners have a strong inclination toward noninvasive methods for arterial blood pressure (ABP) estimation. PPG signals, derived from changes in blood volume in peripheral blood vessels, particularly at the fingertip emerge as a promising method for noninvasive blood pressure measurement. PPG signals are obtained from a

T. Mladenovska (✉) · N. Ackovska · M. Kostoska · B. Koteska · K. T. Dineva · A. M. Bogdanova
Faculty of Computer Science and Engineering, Ss. Cyril and Methodius University, Skopje, North Macedonia
e-mail: mladenovska.teodora@students.finki.ukim.mk; nevena.ackovska@finki.ukim.mk; magdalena.kostoska@finki.ukim.mk; bojana.koteska@finki.ukim.mk; katarina.trojacanec@finki.ukim.mk; ana.madevska.bogdanova@finki.ukim.mk



Ö. M. Gül et al. (eds.), *7th EAI International Conference on Robotic Sensor Networks*, EAI/Springer Innovations in Communication and Computing,
https://doi.org/10.1007/978-3-031-64495-5_13

visible light (LED) pulse oximeter that is applied to the skin and measures the micro-variations of the intensity of light transmitted or reflected via a photo-diode [2]. Recent developments in signal processing and machine learning allowed scientists to create algorithms for SBP and DBP estimation on the basis of PPG signals. In general, these algorithms function by examining the PPG waveform. Afterward, features that exhibit a correlation with blood pressure are extracted, and they can be used to train machine learning models with the ultimate goal of predicting blood pressure from PPG signals [3, 4]. Instead of focusing only on predicting SBP and DBP values, the main goal of this study is to estimate the arterial blood pressure (ABP) waveform. The objective is to extract blood pressure measurements from the waveform, where the systolic blood pressure is defined based on the highest value, and the diastolic blood pressure based on the lowest value.

The rest of the chapter is organized as follows: Sect. 2 provides insights into related research. In Sect. 3, the materials and methods used for our study are explained. The evaluation metrics applied for evaluating the models' performances are presented in Sect. 4. Section 5 presents and discusses the models' outcomes, and Sect. 6 concludes the paper, summarizing key findings and their broader implications.

## 2 Related Work

The estimate of the arterial blood pressure (ABP) waveform has been the subject of several research articles. In [5], referred to as PPG2ABP, two deep learning models were employed to estimate the ABP waveform. Initially, an approximation network, a one-dimensional U-Net network, takes a PPG signal as input and is used to approximate the waveform. Subsequently, a refinement network was applied to improve the estimated ABP waveforms, incorporating a 1D MultiResUNet model for enhancement. A different approach was taken in another work [6], predicting the ABP waveform through the BP-Net that was developed as an extension of the U-Net.

Comparison between two deep convolution autoencoders, LeNet-5 and U-Net, for predicting ABP waveforms was conducted in [7] using cross-validation (CV). The results indicated that U-Net outperformed alternative techniques for systolic blood pressure (SBP) values, while LeNet-5 showed slight superiority in predicting diastolic blood pressure (DBP) values. Further optimization of the ensemble of CV models was achieved through a deep convolutional autoencoder (GDCAE) based on a genetic algorithm, demonstrating superior performance compared to LeNet-5 and U-Net. However, the computational inefficiency of combining two deep learning algorithms with numerous parameters was noted, and no optimal model for predicting the resulting ABP waveform was generated. In a recent study [8], the suggestion was made to use PPG data as input for a cycle generative adversarial network (CycleGAN) in predicting ABP waveforms. While the majority of publications leaned toward encoder–decoder technology, this particular study adopted an innovative approach, employing a generator and a discriminator network for waveform estimation.

# 3 Materials and Methods

## *3.1 Dataset*

The MIMIC-III dataset is a publicly accessible repository of health-related data, containing information from over 40,000 patients [9]. In this chapter, our analysis is focused on a subset of the MIMIC-III dataset, specifically on 77 patients. The data was sampled at a rate of 125 samples per second. Our dataset selection criteria involved specifically choosing records where PPG and ABP signals are within the time range of 4 seconds to 6 minutes, ensuring complete coverage of the entire time window without any missing values. Additionally, data filtering was applied to include cases where the correlation between ABP and PPG signals exceeds 0.92. To prepare the input vectors, we also use the difference between two adjacent values, and the output from the models is a raw ABP signal.

## *3.2 Preprocessing*

One of the major issues impeding the full utilization of available biological databases is the preservation of these data without regard for their quality. The data cleaning was done by employing several preprocessing approaches often employed when working with time series data and biological signals. The phases of data preparation included: filtering and normalization, segmentation and correlation, and signal alignment (between PPG and APB).

### 3.2.1 Filtering and Normalization

Motion artifacts (MA), which appear due to the design of PPG sensors, can cause distortions in signal quality. One of the most common causes of MA is related to changes in blood flow resulting from light penetration between the sensor and the skin. Another is the physical activity that alters the sensor's placement on the skin [10].

In this research, we performed data filtering in a way that PPG signals were filtered in first place with the Python Neurokit tool at a sampling rate of 125 Hz. Once the data is cleaned, the MinMaxScaler is utilized for scaling.

#### 3.2.2 Signal Alignment

ABP and PPG measurements are frequently provided without time alignment, and out of phase because of disparate methodologies used and data collection from different body parts (such as arm vs. wrist) [11, 12]. The alignment process has been done for all records using cross-correlation. The formula is given below.

$$g(\Delta t) = \sum \mathbf{ABP}(t) \times \mathbf{PPG}(t + \Delta t)$$

#### 3.2.3 Segmentation and Correlation

Pearson's correlation coefficient is employed to assess the morphological similarity between PPG and ABP. The coefficient's value, denoted as "r," ranges from−1 to +1. A value close to −1 or +1 indicates strong negative or positive similarity, respectively, while a value close to zero suggests a lack of significant similarity [13]. The correlation coefficient is defined as:

$$r = \frac{n\sum XY - \sum X \sum Y}{\sqrt{n\sum X^2 - (\sum X)^2} - \sqrt{n\sum Y^2 - (\sum Y)^2}}.$$

According to the existing literature, PPG and ABP signals are considered morphologically associated if the correlation factor (r) exceeds 0.92 [13]. Consequently, signals with correlations less than 0.92 are excluded. Additionally, the maximum and minimum values of the ABP signals are determined. If the maximum value exceeds 220 (SBP $>$ 220) and the minimum value is below 40 (DBP $<$ 40), the signal is deemed outside the scope of our work and is rejected. Because of the pre-segmentation partitioning of sets, patients included in the training set are not present in the testing and validation sets, and conversely. The division of the dataset is done with an allocation of 8% for testing, 7% for validation, and 85% for training. Each patient may have different amounts of recordings, and each record's length may vary.

### 3.3 *Model Structure*

This paper employs two model types: a deep neural network and an encoder–decoder model. The deep neural network specifically utilized is a Gated Recurrent Unit (GRU), which belongs to the category of Recurrent Neural Networks (RNNs).

The second type of the used network is an encoder–decoder model. The encoder is processing the input sequence and extracting valuable information from it. This input sequence could be in the form of text, images, or any other structured data.

It utilizes layers of neural networks, including recurrent neural networks (RNNs), convolutional neural networks (CNNs), or transformer models, to sequentially or spatially analyze the input. As it processes the input, the encoder produces a fixed-size representation known as the "context vector," which captures the important features and information from the input data. The decoder utilizes the context vector generated by the encoder to generate the intended output sequence. Similar to the encoder, the decoder also utilizes layers of neural networks, frequently configured as recurrent neural networks (RNNs), long short-term memory networks (LSTMs), or transformer decoders. The decoder generates the output sequence step by step, often in an autoregressive manner [14]. The structures of both networks were contingent on the input vector sizes and underwent training for 50 epochs with a batch size of 32, utilizing the Adam optimizer with a learning rate set at 0.0001 and incorporating checkpoints for early stopping and saving the optimal model.

## 4 Evaluation Metrics

The assessment technique employs the evaluation metrics referred to as Mean Absolute Error (MAE), Mean Squared Error (MSE), and R-squared (r2_score) [15]. The Mean Squared Error (MSE) and the Mean Absolute Error (MAE) are often used loss functions in regression analysis. The MSE utilizes the L2 loss function, while the MAE employs the L1 loss function.

$$\mathrm{MSE} = \frac{\sum_{i=1}^{n}(y_i - \hat{y}_i)^2}{n}$$

$$\mathrm{MAE} = \frac{\sum_{i=1}^{n}|y_i - \hat{y}_i|}{n}$$

In addition to these two metrics, a frequently used metric for regression problems is r2_score, indicating how well the independent variable(s) can explain the variation of a dependent variable. It is calculated according to the following:

$$R^2 = 1 - \frac{SS_{Regression}}{SS_{Total}} = 1 - \frac{\sum_i(y_i-\hat{y}_i)^2}{\sum_i(y_i-\bar{y})^2}$$

In our experiments, we employ Mean Squared Error (MSE) as the model's error function. MSE penalizes errors by their squared values, meaning that it assigns a greater penalty to outliers. This emphasis on squaring the errors leads to a gradual gradient for small errors.

## 5 Results and Discussion

After completing the training process for the model, the evaluation is performed using the designated test dataset, leading to the following results.

Table 1 shows the MAE, MSE, and R2 scores for the ABP waveform using different models with different window sizes. The window length of 175 signal points corresponds to a window size of 1s and the window length of 512 signal points—a window size of 4s. From the above table, we can conclude that the best result for predicting the ABP waveform is obtained using a GRU network with a window size of 4s. After evaluating the whole ABP signal, we also want to evaluate the SBP and DBP values. For the evaluation of these values, we are using MAE and MSE, and the results are shown below.

Table 2 shows that the best value for DBP is obtained by using the encoder–decoder model with a window size of 4s, and the best value for SBP is obtained by using a GRU network with a window size of 1s. The outcomes do not reveal significant distinctions among the various models. However, based on the findings, we can draw the conclusion that a window size of 4s is most appropriate for estimating DBP, while a 1s window size is optimal for SBP estimation.

Figures 1 and 2 show the true (blue line) and the predicted (orange line) ABP waveform, using different window sizes. Observing the results, we can see that even if we are using a different window size the prediction line follows the true line, and the predictions from GRU are more compatible compared with the predictions from encoder–decoder model where we can observe some noise.

**Table 1** Evaluation of ABP waveform

| Window size (number of signal points) | 175 | | | 512 | | |
|---|---|---|---|---|---|---|
| Model | mae | mse | r2 | mae | mse | r2 |
| GRU neural network | 11.96 | 201.85 | 0.75 | **11.28** | **185.37** | **0.76** |
| encoder-decoder | 12.63 | 216.68 | 0.73 | 14.54 | 314.21 | 0.61 |

**Table 2** Evaluation of SBP and DBP values

| Window size (number of signal points) | 175 | | | | 512 | | | |
|---|---|---|---|---|---|---|---|---|
| | DBP | | SBP | | DBP | | SBP | |
| metric / model | mae | mse | mae | mse | mae | mse | mae | mse |
| GRU neural network | 11.94 | 175.70 | 10.48 | 158.43 | 9.35 | 111.51 | 12.70 | 222.75 |
| encoder-decoder | 10.16 | 127.32 | 12.30 | 213.22 | 6.52 | 67.30 | 15.73 | 325.40 |

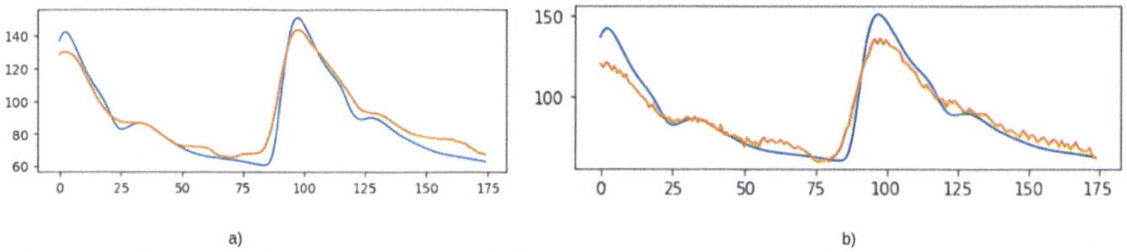


**Fig. 1** True and predicted values for a window size of 1s: (**a**) GRU, (**b**) encoder–decoder architecture

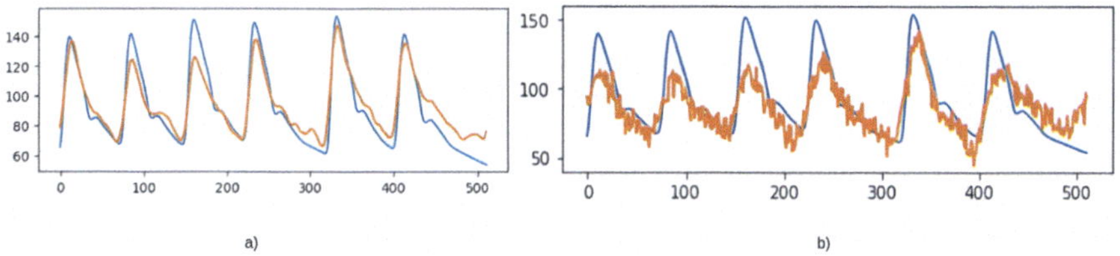


**Fig. 2** True and predicted values for a window size of 4s: (**a**) GRU, (**b**) encoder–decoder architecture

# 6 Conclusion

In this research, our aim was to estimate Arterial Blood Pressure (ABP) waveforms using our proposed models based on Photoplethysmogram (PPG) signal data. These models also estimate Blood Pressure (by extracting the scalars SBP, DBP, and MAP). In this manner, they differ from numerous existing models that focus only on predicting SBP and DBP values as singular scalars from input PPG/ECG signals.

The estimated ABP waveform has the potential for extensive use in medical analysis of cardiovascular irregularities, providing an alternative to conventional invasive medical procedures. With our new models, ABP waveform can now be approximated using noninvasive PPG signals obtained through standard PPG sensors.

Alongside, unlike several studies that employed ECG signals in addition to the PPG signals for ABP waveform estimation, we opted to exclusively handle PPG signals while still achieving encouraging estimate results.

In the future, we will focus on increasing the model performance, enriching the dataset with additional data, and experimenting with other networks and models.

**Acknowledgments** This research was made feasible through the assistance provided by the "Smart Patch for Life Support Systems," NATO project G5825 SP4LIFE.

## References

1. Kara Rogers: Blood pressure. Retrieved: Mar, 2023. https://www.britannica.com/science/blood-pressure
2. Susha Cheriyedath: Photoplethysmography (PPG). Retrieved: Mar, 2023. https://www.news-medical.net/health/Photoplethysmography-(PPG).aspx
3. Harfiya, L.N., Chang, C.C., Li, Y.H.: Continuous blood pressure estimation using exclusively photopletysmography by lstm-based signal-to-signal translation. National Library of Medicine, National Center for Biotechnology Information (2021) https://doi.org/10.3390/s21092952
4. Tazarv, A., Levorato, M.: A deep learning approach to predict blood pressure from PPG signals. CoRR **abs/2108.00099**, 2108.00099 (2021)
5. Ibtehaz, N., Mahmud, S., Chowdhury, M.E.H., Khandakar, A., Ayari, M.A., Tahir, A., Rahman, M.S.: PPG2ABP: Translating Photoplethysmogram (PPG) Signals to Arterial Blood Pressure (ABP) Waveforms using Fully Convolutional Neural Networks (2022)
6. Vardhan, K.R., Vedanth, S., Poojah, G., Abhishek, K., Kumar, M.N., Vijayaraghavan, V.: BP-Net: efficient deep learning for continuous arterial blood pressure estimation using photoplethysmogram. In: 2021 20th IEEE International Conference on Machine Learning and Applications (ICMLA) (2021)
7. Sadrawi, M., Lin, Y.-T., Lin, C.-H., Mathunjwa, B., Fan, S.-Z., Abbod, M.F., Shieh, J.-S.: Genetic deep convolutional autoencoder applied for generative continuous arterial blood pressure via photoplethysmography. Sensors **20**(14) (2020) https://doi.org/10.3390/s20143829
8. Mehrabadi, M.A., Aqajari, S.A.H., Zargari, A.H.A., Dutt, N., Rahmani, A.M.: Novel blood pressure waveform reconstruction from photoplethysmography using cycle generative adversarial networks. In: 2022 44th Annual International Conference of the IEEE Engineering in Medicine & Biology Society (EMBC), pp. 1906–1909. IEEE (2022)
9. Johnson, A.E.W., Pollard, T.J., Shen, L., Lehman, L.-w.H., Feng, M., Ghassemi, M., Moody, B., Szolovits, P., Celi, L.A., Mark, R.G.: Mimic-iii, a freely accessible critical care database. National Library of Medicine, National Center for Biotechnology Information (2016) https://doi.org/10.1038/sdata.2016.35
10. Elgendi, M.: On the analysis of fingertip photoplethysmogram signals. Current Cardiol. Rev. **8**, 14–25 (2012) https://doi.org/10.2174/157340312801215782
11. Harfiya, L.N., Chang, C.-C., Li, Y.-H.: Continuous blood pressure estimation using exclusively photopletysmography by lstm-based signal-to-signal translation. Sensors **21**(9), 2952 (2021)
12. Attivissimo, F., De Palma, L., Di Nisio, A., Scarpetta, M., Lanzolla, A.M.L.: Photoplethysmography signal wavelet enhancement and novel features selection for non-invasive cuff-less blood pressure monitoring. National Library of Medicine, National Center for Biotechnology Information (2023) https://doi.org/10.3390/s23042321
13. Aguilar, G.M., Howard, N., Abbott, D., Lim, K., Ward, R., Elgendi, M.: Can photoplethysmography replace arterial blood pressure in the assessment of blood pressure? J. Clin. Med. **7**, 316 (2018) https://doi.org/10.3390/jcm7100316
14. Nadeem: Encoder decoder. Retrieved: Mar, 2023. https://medium.com/analytics-vidhya/encoders-decoders-sequence-to-sequence-architecture-5644efbb3392
15. Brownlee, J.: Regression Metrics for Machine Learning. Retrieved: Mar, 2023. https://machinelearningmastery.com/regression-metrics-for-machine-learning/

# Explainable AI for Discovering Disease Biomarkers: A Survey

**Slave Temkov**

## 1 Introduction

The pursuit for discovering biomarkers—molecular signatures indicative of physiological or pathological states—cultivates the advancement of precision medicine. Biomarkers play a substantial role in disease diagnosis, prognosis, and therapeutic monitoring, enabling a personalized approach to patient care. However, the task of biomarker discovery is a complex venture, often requiring the analysis of high-dimensional genomic data to unearth meaningful molecular signatures [1]. This quest has been significantly fostered by the emergence of sensitive omics platforms, which enable the measurement of a myriad of biological molecules at an unprecedented scale [6].

The integration of artificial intelligence (AI) and machine learning (ML) has heralded a new epoch in biomarker discovery. These computational paradigms, adept at handling high-dimensional data, have emerged as powerful tools for identifying potential biomarkers across a diverse spectrum of diseases. Machine learning models, with their ability to learn complex patterns in data, have been pivotal in advancing the frontier of biomarker discovery. However, conventional machine learning models often suffer from a lack of interpretability, rendering them as "black-box" models. This black-box nature precludes a mechanistic understanding of the biomarker-disease relationships, a caveat that is particularly concerning in the clinical context where interpretability and transparency are imperative [7].

S. Temkov (✉)
Faculty of Computer Science and Engineering, University ss. Cyril and Methodius, Skopje, North Macedonia
e-mail: slave.temkov@finki.ukim.mk
http://www.finki.ukim.mk



Ö. M. Gül et al. (eds.), *7th EAI International Conference on Robotic Sensor Networks*, EAI/Springer Innovations in Communication and Computing,
https://doi.org/10.1007/978-3-031-64495-5_14

The introduction of explainable artificial intelligence (XAI) addresses the interpretability lacuna in machine learning models. XAI techniques furnish transparency to machine learning predictions, providing mechanistic insights into the discovered biomarkers. By rendering the models interpretable, XAI paves the way for a deeper understanding of the molecular mechanisms underlying diseases, thereby fostering a more robust validation of the discovered biomarkers. The combination of XAI with machine learning models in biomarker discovery has resulted in a harmonious synergy, where not only the identification of biomarkers is accelerated but also the underlying biological narratives are unraveled [10].

This survey presents an overview of recent state-of-the-art and emerging XAI applications in the realm of biomarker discovery for various diseases, providing a snapshot of the latest advancements in this research area. We will traverse through the critical facets of leveraging XAI for efficient disease biomarker discovery, encompassing the landscape of the data utilized and its requisite preprocessing, the AI and ML models employed and their application in the task, the diverse explainability methods used, and the attainable results. The overarching goal is to highlight the transformative potential of XAI in propelling precision medicine forward, offering a perspective for innovative diagnostic and therapeutic strategies.

## 2 Data: Landscape and Preprocessing

Regardless of what kind of AI and ML task, collecting all the necessary data in a right format is always crucial in the first steps [8]. In the quest for biomarker discovery, the nature and quality of data employed play a pivotal role as well. The corpus of literature surveyed includes a diverse range of data types and sources leveraged for unearthing biomarkers across many diseases. Central to this endeavor is the use of genomic data, which encodes the molecular narrative pivotal for understanding disease mechanisms. As the task of finding gene markers for diseases comes down to finding the differentially expressed genes (DEGs) [7], the base data that is being used most commonly is gene expression data in the format of microarrays. However, this base data is sometimes accompanied by other more specific features, depending on the illness that is being targeted, as well as the particular task that is being performed. The exploration of genomic data, as seen in the papers focused on cancers and Alzheimer's disease, underscores the significance of high-resolution data in biomarker discovery [2, 9].

The diseases that have been studied using XAI in the surveyed papers—mostly cancerous diseases and in one case Alzheimer's—are usually of this nature, where the genetic significance is prevalent.

The advent of sensitive omics platforms has catalyzed the acquisition of high-dimensional biological data, providing researchers with a rich assortment of information for biomarker discovery [6]. These platforms enable the measurement of a myriad of biological molecules at an unprecedented scale, significantly improving the prospects for robust and reproducible biomarker discovery [3]. The data harvested from such platforms often comprise gene expression data, genomic sequences, methylation data, and more, each harboring potential biomarkers for a wide spectrum of diseases.

The preprocessing of this vast and intricate data is a crucial step toward ensuring the accuracy and reproducibility of the findings. Among the surveyed literature, various preprocessing techniques have been employed to distill meaningful insights from the data. For instance, the use of principal component analysis (PCA) to eliminate redundant genes and reduce data dimensionality is highlighted in the paper exploring biomarker discovery for liver cancer [5]. Such dimensionality reduction techniques are imperative for managing the computational complexity and improving the signal-to-noise ratio in the data. In the scope of biomarker discovery, this way we can omit genes with expression that is very similar in individuals that have or do not have the disease.

Furthermore, the papers also highlighted the importance of data normalization and filtering as pivotal preprocessing steps. Normalization ensures a uniform scale for the features, facilitating a more accurate analysis, while filtering helps in eliminating noise from the data, honing in on the most significant features for biomarker discovery.

Moreover, the use of machine learning and XAI methods for feature selection is a recurrent theme across the surveyed papers. These methods not only help in identifying potential biomarkers but also provide a mechanistic understanding of the relationships between the features and the disease phenotypes. The use of explainable AI methods offers a deeper understanding of the molecular mechanisms underlying the diseases, thereby aiding in the validation of the discovered biomarkers before proceeding to costly and time-consuming validation studies [7].

The diversity in data types and preprocessing techniques showcased in the surveyed literature underscores the multifaceted nature of biomarker discovery. As the field advances, the confluence of robust preprocessing techniques and XAI is poised to play a pivotal role in harnessing the full potential of the rich genomic data for biomarker discovery, thereby advancing the frontier of precision medicine.

A significant limitation in the use of XAI for biomarker discovery is the availability and quality of data. Biomarker discovery often relies on large, high-quality datasets, which may not always be available or may contain biases that can affect the model's performance and interpretability. Furthermore, integrating different types of data (e.g., genomic, proteomic, clinical data) into a cohesive model that can be effectively interpreted by XAI is a complex task that requires careful consideration of data compatibility and preprocessing techniques.

In summary, the data landscape and preprocessing techniques employed in biomarker discovery are as diverse as the diseases they aim to elucidate. The meticulous preprocessing of high-dimensional data, coupled with the power of XAI, sets a robust foundation for unearthing biomarkers that could significantly impact diagnostic and therapeutic strategies across a myriad of diseases.

## 3 Methods and Results

The methodological approaches adopted in the quest for biomarker discovery, as explained in the surveyed literature, underscore a harmonious interplay between advanced machine learning algorithms and explainable artificial intelligence (XAI) techniques. This synergy is instrumental in not only identifying potential biomarkers but also in furnishing a deeper understanding of the underlying molecular mechanisms.

Central to the methodological discourse is the use of machine learning classification models. These models are adept at handling the high-dimensional data typical in biomarker discovery, effectively distinguishing between signal and noise. For instance, the employment of classification models like random forest (RF) and the utilization of dimensionality reduction techniques like principal component analysis (PCA) have been pivotal in identifying genes associated with hepatocellular carcinoma (HCC) [5]. In simpler models such as decision trees, RF, and so on, there are already intrinsic methods for determining feature importance that can be used directly to identify the biomarker genes of a disease, depending on the nature of the data and the prediction task.

However, the black-box nature of conventional machine learning models often excludes a mechanistic understanding of the biomarker-disease relationships. This is where explainable AI (XAI) techniques bridge the interpretability gap. Techniques like Local Interpretable Model-agnostic Explanations (LIME) and Shapley analysis provide post hoc explanations of the machine learning predictions, furnishing mechanistic insights into the discovered biomarkers [2]. The case exploring biomarker discovery in ovarian cancer, for instance, delineates how Shapley analysis can be harnessed for explicating the contributions of individual features in the classification task. The LIME analysis is specifically very useful when it comes to use cases of personalized treatments.

The results emanating from these methodological paradigms are nothing short of transformative. The identification of critical biomarkers such as the ORAI2 gene in Alzheimer's disease and the elucidation of genetic biomarkers in various cancers exemplify the potential of these methods in advancing the frontier of precision medicine [9]. Furthermore, the application of XAI in the realm of infectious diseases like COVID-19 has also been explored, demonstrating the versatility and imperative of XAI in biomarker discovery across a diverse disease spectrum [11].

Moreover, the case study delving into biomarker discovery for rheumatoid arthritis illustrates the accessibility of AI and machine learning tools, demonstrating how these advanced methodologies can be harnessed for disease interrogation and response mechanism analysis [7].

Viewing most of the surveyed papers, we can infer a general design for the framework of a solution when it comes to biomarker discovery using XAI (Fig. 1).

There is one specific case that examines this task from a different point of view—the case of using XAI for precision medicine in myeloid leukemia [4]. In this paper, the last part of this whole precision medicine is being pursued—the association

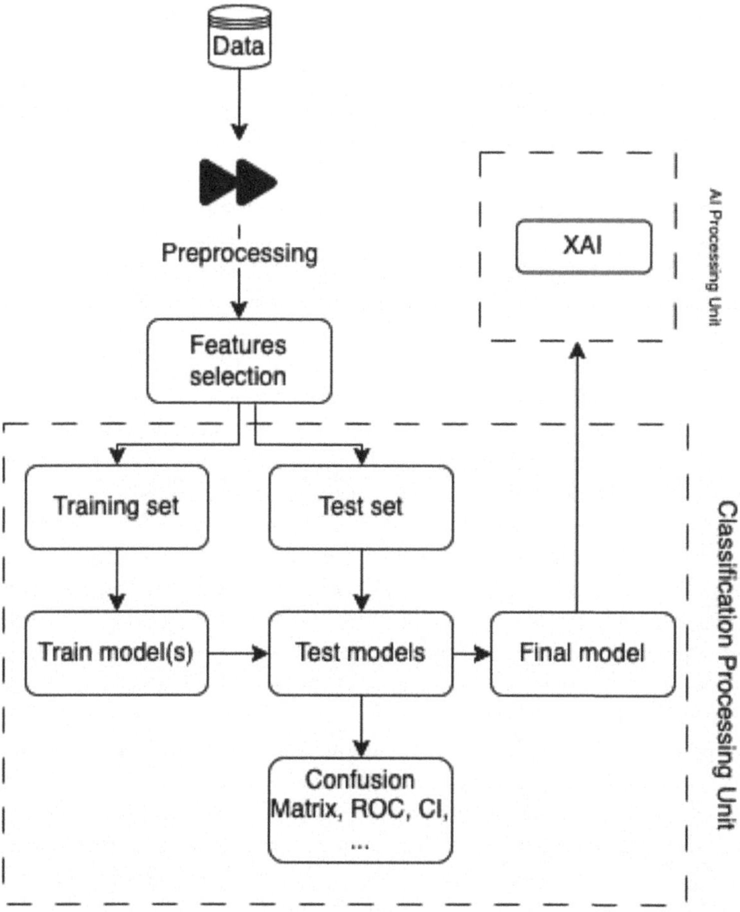


**Fig. 1** A general framework for biomarker discovery through microarray analysis of gene expression data [5]

of drug screening with genomic events. This is done through a proposed novel method—multidimensional module optimization (MOM). This method includes two steps of data preprocessing, and in the last step, it utilizes mixed integer linear programming (MILP) optimization for predicting the combination of drugs and biomarkers that optimize patient response to treatment. The predictions from MOM are interpretable and robust, making it a suitable tool for clinical applications. It offers an optimized treatment strategy for each patient, taking into account drug toxicity and ensuring that each patient receives the best possible treatment. By associating drugs with genetic events, it also facilitates the identification of targeted treatments based on patient genotype, potentially improving treatment efficacy. Last but not least, the method is designed to be easily understood by clinicians and can be implemented in clinical practice, resembling a decision tree-based approach.

This feature is crucial for its adoption in healthcare settings. It is an interesting novel method that can be applied for various use cases—other types of cancer or diseases where drug response is influenced by genetic biomarkers; to identify potential drug candidates and their effectiveness based on specific genetic profiles in the process of drug development and screening; in the design and stratification of patients in clinical trials to make them more targeted and efficient, etc. The method could be expanded to include not just genetic data but also other types of OMICS data, clinical data, and patient-reported outcomes, providing a more holistic view of patient health and treatment response. The great potential in it is that it tackles the final goal of all these analysis and biomarker discovery—precise and targeted medicine.

The clinical validation of the biomarkers identified by XAI models is crucial. This involves extensive testing and confirmation through independent studies and clinical trials, which can be time-consuming and resource-intensive. It's also important to ensure that the biomarkers identified are clinically relevant and can be translated into practical diagnostic or therapeutic applications.

The discussions in the surveyed papers also touch upon the potential pitfalls and challenges. The propensity of machine learning models to overfit, especially in scenarios with limited data, and the high likelihood of false discovery are concerns that necessitate a meticulous validation of the discovered biomarkers. Therefore, it is really important to evaluate the performance of the trained models with various metrics, before employing any explainability methods for finding biomarkers and drawing conclusions.

In summary, the methodological paradigms adopted in the surveyed literature underscore the multifaceted nature of biomarker discovery. The fusion of machine learning and XAI not only accelerates the pace of biomarker discovery but also furnishes a robust platform for a deeper understanding of the complex disease mechanisms. The resultant discoveries, as elucidated in the surveyed papers, hold significant promise in advancing diagnostic and therapeutic strategies, thereby propelling the field of precision medicine into a new era of innovation and understanding.

## 4 Conclusion

The quest of biomarker discovery, as portrayed in the surveyed literature, underscores a vibrant and promising intersection between explainable artificial intelligence (XAI) and machine learning methodologies. The entwined narrative of these advanced computational paradigms with the intricate tapestry of genomic data heralds a new era in the realm of precision medicine. This survey has endeavored to traverse the methodological, data, and analytical landscapes that underpin biomarker discovery across a diverse spectrum of diseases.

The salient takeaway is the transformative potential of XAI in demystifying the complex biological systems, thereby fostering a deeper understanding of disease mechanisms. Through a meticulous dissection of the various methodological paradigms, this survey has elucidated how XAI furnishes transparency and interpretability to machine learning models, a virtue that is pivotal for validating the discovered biomarkers before embarking on costly and time-consuming validation studies [7].

Moreover, the diversity in data types, preprocessing techniques, and the resultant biomarker discoveries underscore the multifaceted nature of biomarker research. The surveyed literature epitomizes the versatility of XAI in biomarker discovery across a gamut of diseases, from cancers to neurological and infectious diseases.

Furthermore, the methodological narratives discussed herein also illuminate the challenges that pervade the field, such as overfitting, high likelihood of false discovery, and the imperative for robust validation frameworks. These challenges beckon a continued exploration and refinement of the existing methodologies and validation paradigms.

As the frontier of biomarker discovery continues to expand, the harmonious fusion of XAI and machine learning is poised to play a pivotal role. The resultant discoveries, as elucidated in the surveyed literature, not only hold significant promise in advancing diagnostic and therapeutic strategies but also in propelling the field of precision medicine into a new era of innovation and understanding.

The voyage of biomarker discovery is far from over; the surveyed literature heralds just the beginning. The confluence of robust data, advanced computational methodologies, and the insatiable quest for understanding the molecular underpinnings of diseases sets a promising trajectory for future research endeavors in this interdisciplinary domain. Through a continued exploration of XAI and machine learning methodologies, the field is poised for uncovering biomarkers that could significantly impact the global healthcare landscape, heralding a new epoch of personalized medicine and patient care.

## References


1. Aronson, J.K., Ferner, R.E.: Biomarkers—a general review. Current Prot. Pharmacol. **76**(1) (2017). https://doi.org/10.1002/cpph.19
2. Chakraborty, D., Gutierrez-Chakraborty, E., Rodriguez-Aguayo, C., Başağaoğlu, H., Lopez-Berestein, G., Amero, P.: Discovering genetic biomarkers for targeted cancer therapeutics with eXplainable AI (2023). https://doi.org/10.1101/2023.07.24.550346
3. Fiscal, L.L., Jennebauffe, C., Bruyneel, M., Ris, L., Lefebvre, L., Siebert, X., Gosselin, B.: Explainable ai for EEG biomarkers identification in obstructive sleep apnea severity scoring task. In: 2023 11th International IEEE/EMBS Conference on Neural Engineering (NER). pp. 1–6 (2023). https://doi.org/10.1109/NER52421.2023.10123795
4. Gimeno, M., José-Enériz, E.S., Villar, S., Agirre, X., Prosper, F., Rubio, A., Carazo, F.: Explainable artificial intelligence for precision medicine in acute myeloid leukemia. Front. Immunol. **13** (2022). https://doi.org/10.3389/fimmu.2022.977358
5. Hasan, M.E., Mostafa, F., Hossain, M.S., Loftin, J.: Machine-learning classification models to predict liver cancer with explainable AI to discover associated genes. AppliedMath **3**(2), 417–445 (2023). https://doi.org/10.3390/appliedmath3020022

6. Lauwen, S., de Jong, E.K., Lefeber, D.J., den Hollander, A.I.: Omics biomarkers in ophthalmology. Investigat. Opthalmol. Visual Sci. **58**(6), BIO88 (2017). https://doi.org/10.1167/iovs.17-21809
7. Ng, S., Masarone, S., Watson, D., Barnes, M.R.: The benefits and pitfalls of machine learning for biomarker discovery. Cell Tissue Res. **394**(1), 17–31 (2023). https://doi.org/10.1007/s00441-023-03816-z
8. Roh, Y., Heo, G., Whang, S.E.: A survey on data collection for machine learning: a big data - AI integration perspective. IEEE Trans. Knowl. Data Eng. **33**(4), 1328–1347 (2021). https://doi.org/10.1109/TKDE.2019.2946162
9. Sekaran, K., Alsamman, A.M., Doss, C.G.P., Zayed, H.: Bioinformatics investigation on blood-based gene expressions of alzheimer's disease revealed ORAI2 gene biomarker susceptibility: an explainable artificial intelligence-based approach. Metabolic Brain Disease **38**(4), 1297–1310 (2023). https://doi.org/10.1007/s11011-023-01171-0
10. Shaban-Nejad, A., Michalowski, M., Brownstein, J.S., Buckeridge, D.L.: Guest editorial explainable AI: towards fairness, accountability, transparency and trust in healthcare. IEEE J. Biomed. Health Inf. **25**(7), 2374–2375 (2021). https://doi.org/10.1109/JBHI.2021.3088832
11. Yagin, F.H., Cicek, İ.B., Alkhateeb, A., Yagin, B., Colak, C., Azzeh, M., Akbulut, S.: Explainable artificial intelligence model for identifying COVID-19 gene biomarkers. Comput. Biol. Med. **154**, 106619 (2023). https://doi.org/10.1016/j.compbiomed.2023.106619

# Part V
# Cybersecurity

# Analytical Approach for Mitigating Phishing Attempts

**Sara Alshehhi, Hessa Aldhaheri, Ameera Alzaabi, Sara Alhammadi, Alanod Alhosani, and Artur Koci**

## 1 Introduction

I. In this digital age, we often receive web links [universal resource locator (URL)] on messages on various platforms like email, instant messaging, and social media. However, there is a large amount of untrustworthy links and spams that target such communication tools. Getting access to such malicious links could leave the user vulnerable to malware and privacy issues. Both organizations and individuals would benefit from a method to avoid this. The need for a solution that can protect several categories of users from falling victim to unsafe URL links received as instant messages or email is pressing. The absence of a reliable and user-friendly system that can assess the safety of URL links contained in messages is a major issue, and existing solutions are either limited in scope or require significant technical expertise. The proposed solution will include multiple security features for real-time threat detection and URL link analysis and will be user-friendly and compatible with popular messaging platforms and email services. The research will address concerns related to the required features and security measures such a tool could provide, the user-friendliness of the solution, testing for reliability and effectiveness, and challenges related to time, resources, evolving threats, and data privacy regulations. Lastly, the aim of this research is to design and develop an application that verifies the safety of URL links received in messages before accessing it and provides a secure and reliable online experience for users.

II. The proposed solution will serve as a stepping stone to implementing a fully smart and automated system for handling online threats. We will look at

S. Alshehhi · H. Aldhaheri · A. Alzaabi · S. Alhammadi · A. Alhosani · A. Koci (✉)
Abu Dhabi Polytechnic, Abu Dhabi, UAE
e-mail: artur.koci@actvet.gov.ae; Artur.koci@adpoly.ac.ae



Ö. M. Gül et al. (eds.), *7th EAI International Conference on Robotic Sensor Networks*, EAI/Springer Innovations in Communication and Computing,
https://doi.org/10.1007/978-3-031-64495-5_15

machine learning as a possible implementation of this issue and collect data that can be used for future systems. As online threats continue to evolve, it is crucial to have reliable and effective solutions to keep users safe. This research will contribute to addressing this pressing need and provide a framework for developing future solutions in this field.

III. We used a systemic test-driven methodology to conduct this research. By iterating the system using the compiled data from malicious websites, we managed to reach a point of functional accuracy in our design. After creating the initial algorithm, we ran tests by removing links from the blacklist and running them through the analysis to see if the system is able to reidentify them as malicious links. Throughout multiple iterations, we managed to reach around 72% accuracy with new links, and 90% accuracy with links that have a close match in the database.

### *1.1 Objectives*

Our main objectives for this research are:

- To design and develop an application that verifies the safety of URL links contained in messages to provide a secure and reliable online experience for users
- To create a comprehensive and user-friendly system that can assess the safety of URL links in messages, protecting users from potential threats without requiring significant technical expertise
- To implement multiple security features in the developed application to provide real-time threat detection and URL link analysis
- To test the reliability and effectiveness of the developed algorithm and evaluate its performance using metrics

We may face several constraints that must be considered during development:

- The constantly evolving nature of online threats, which means that the application must be updated regularly to keep up with new threats.
- Adhering to data privacy regulations. This is to ensure that user data is protected and not shared with third parties.
- Balancing these constraints and delivering a secure, reliable, and user-friendly solution is a major challenge for our team.

## 2 Prior Research

### *2.1 Related Work*

With the immense increase in the use of the Internet by users of different categories of age, phishing attempts are becoming an obstacle for many business solutions.

Such an obstacle has been drawing the attention of many authors to conduct research and propose different solutions to phishing attempts. Throughout the literature, there are multiple approaches found that aim to mitigate the issue of malicious links. In this section, we analyze the main methods proposed for mitigating phishing attempts.

One of the most well-known solutions proposed by [1, 2] is called keyword search that involves searching through the web link for keywords that identify a malicious website. It does not require user intervention and is simpler and easier than AI methods, but it is vulnerable to false positives and false negatives and can be computationally heavy. There are also website masking techniques that can hide malicious content.

Another approach proposed by [3] is user flagging, which relies on users to flag malicious websites, preventing other users from accessing them. It is self-updating and easier to use for nontechnical users but relies on having an active and technically savvy user base and is very vulnerable to false negatives.

The two-factor authentication proposed by [4] is not an ideal solution [4] since most non-tech-savvy users would ignore it, and forcing such a method on them seems to be a cause for annoyance. On the other hand, user flagging is even more demanding in technical knowledge from users and less useful on its own. Looking into such solutions and their weaknesses, our approach is to combine user flagging with keyword search, covering each other's weaknesses. Keyword search will catch any website that specifically targets inexperienced users while at the same time more experienced users will submit links that pass by the filtering system to be flagged. With time, the database will contain enough data to protect more users than a two-factor authentication system can.

## 2.2 *Phishing Detection Algorithms and Techniques*

Various techniques are used for detecting malicious links to protect users. In the twenty-first century, banks have adopted measures such as one-time passwords and visual cryptography to authenticate both clients and the employees while accessing the bank's online services. In their research, Nanda and Gupta's approach [5] asks the user to first enter their username and password, and after receiving a one-time password (OTP) on a different phone that they will use to uniquely identify themselves and get access to the online bank services. For any subsequent transactions or changes, the user must provide their fingerprint for authentication. The division of fingerprints into two shares further adds to the security of the system. The idea is that even if the password and username were leaked by a phishing attack, it would still be unlikely that the perpetrator would manage to replicate the fingerprints or the OTP [5].

Deceptive phishing is a significant issue in instant messaging, where much sensitive and personal information is disclosed through socially engineered text messages. Unlike static phishing, which can be dealt with using blacklists, dynamic phishing, where someone pretends to be trustworthy on message apps, is a much harder issue to solve. A solution to this problem is proposed by Ali and Rajamanji [1], where they use data mining techniques to analyze known phishing attacks and extract phishing words from a conversation, and when it reaches a user-defined limit, it sends an alert.

Sharma, Dash, and Ansari [6] discuss the use of AI and machine learning anti-phishing algorithms to detect phishing attacks and cite Google's algorithm for spam detection on Gmail. They conclude that these algorithms are effective against this type of attack [7].

In a phishing detection system based on machine learning [7], the authors discuss how machine learning can be utilized for phishing detection. The proposed system focuses on URL-based detection, utilizing machine learning, specifically support vector machine (SVM) algorithms. Unlike traditional blacklist-based methods, this approach aims to reduce labor costs and detect phishing attacks in real time, even zero-day attacks that have not yet been added to blacklists [7].

Fuzzy logic is incorporated into the machine learning system to improve accuracy. The system extracts various features, including the similarity parameters between the webpage's URL and URLs within the webpage, subdomains, and paths. Additionally, the system considers the time-to-live of webpages to improve detection accuracy. To alert users when accessing phishing webpages, part of the system is implemented as a browser extension, providing real-time protection. The article also briefly discusses related phishing detection techniques, including content-based and URL-based methods, as well as machine learning approaches. It compares SVM with decision tree and logistic regression algorithms, demonstrating SVM's superior performance [7]. In experiments using a dataset of 5000 phishing pages and 10,000 normal webpages, the system achieved an accuracy of 89.3% in detecting phishing webpages, with a false-positive rate of 6.2%.

However, the article notes that certain phishing pages, particularly those using screenshots of legitimate websites and placing input boxes strategically, might require image analysis for detection [3].

In "an enhanced phishing detection tool using deep learning from URL" [8], the authors present an enhanced phishing detection tool that leverages deep learning techniques to identify potentially harmful phishing content in URLs. The tool aims to address the increasing prevalence of phishing attacks, which pose significant cybersecurity threats to individuals and organizations worldwide. It uses a large dataset of URLs, including both legitimate and phishing websites, to train and test the system. The tool is designed to detect phishing attacks with high accuracy rates and is supported by features like a blacklist of websites and various APIs to reduce time consumption. The authors [8] highlight the critical importance of cybersecurity, given the widespread reliance on computer and Internet systems in various sectors of the global economy.

The tool works by using the technology of deep learning to identify potential phishing threats within URLs. Its workflow begins with the collection and preparation of a comprehensive dataset, which includes both legitimate and phishing websites. Users can modify the model based on their needs by using feature selection. During model training, the system employs various algorithms, such as neural networks and random forests, to learn and recognize patterns indicative of phishing attacks. By leveraging this blend of machine learning techniques and a diverse dataset, the tool demonstrates its ability to detect phishing websites effectively, contributing to enhanced cybersecurity measures in an increasingly digitized world [8].

In a study [9], a hoax detection system was developed using Levenshtein distance to identify email hoaxes. The system evaluates the email by comparing it with a set of known scams. The proposed system also includes a feature that informs recipients of the likelihood that the email is a hoax, distinguishing it from previous systems. The system achieved a high positive predictive value of 0.96; however, it cannot identify genuine emails. The ability of the system to detect hoaxes sent in image format is limited, and the collection of hoax keywords is restricted, highlighting areas where the system can be further improved.

## 3 Proposed Phishing Detection Algorithm Design

The structure of our algorithm is shown in Fig. 1. The user clicks on the received link, and once the link is clicked it will automatically activate the phishing detection tool and the link will be sent to the tool database for investigation. If the link was pre-flagged, it will not open on the browser, and it will be further investigated and analyzed by the keyword search engine. After the investigation from the keyword search engine, if the returned result is flagged link, the link will be blocked and not opened by the browser. In case the link is not flagged in one of the examinations above, it is allowed to open normally from the browser. If a user still believes the link is malicious, they can submit it to the database for review, and it is added to the flagged list if it is found to be malicious.

The functions of the system, as shown in Fig. 1, are as follows.

The first function of the system is that the user receives a link through any messenger app that connects to the system. This means that the system can be integrated into different messenger apps and users can receive links through any of these apps. The integration with different messenger apps makes it more convenient for users to use the system since they do not have to switch to a different app to receive links.

The second function of the system architecture as seen in picture 1 is that when the user clicks on the link for the first time, our app is suggested as a way to open the link. This is a helpful feature as it guides the user to use the system to open the link. This is especially useful for users who may not be familiar with the system or how to use it. Additionally, the ability for the user to choose to make the app the default for opening such links ensures that the system is used consistently for all links.

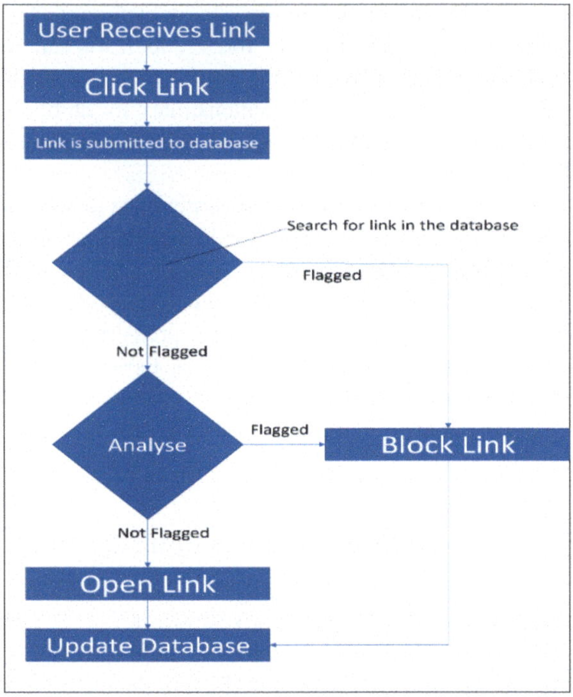


**Fig. 1** System design

The third function of the system is that the link is automatically submitted to the database before it is sent to the browser. This ensures that the link is checked against a database of blocked links before the user is directed to the browser. By submitting the link to the database first, the system can quickly determine whether the link is a phishing link or not and prevent the user from accessing it.

The fourth function of the system is that the link is searched for in the blocked list of the database. If the link is flagged, it is blocked, and the system proceeds to step 7. This step ensures that any links that are already known to be malicious or phishing links are blocked before the user is directed to the browser. This helps prevent users from accidentally accessing malicious links.

The link is analyzed using keyword and text-based analysis. The link is matched with the full database using a Levenshtein distance algorithm, which compares how many changes need to be done on one string to transform it into another. This way we not only make sure the link itself is not flagged but we also guarantee that it does not look similar to other flagged links. Then, we also compare it to a list of popular websites that are fishing targets. These check if the link is intentionally trying to look similar to a real website (e.g., twitter.com could have a phishing website called twiter.com). We also look for common phishing link markers like no SSL certificate and unusual domains. If it was flagged, then it is blocked, proceed to step 7. If it was not flagged, proceed to step 6.

The sixth function of the system is that the app sends the link to the default browsing app if there is one. This means that if the link is not flagged as a phishing link, the user is directed to their default browsing app to view the website. This ensures that the user experience is seamless, and there are no unnecessary interruptions to their browsing experience

The seventh and final function of the system is that if the link is flagged as a phishing link, it is automatically added to the database. This ensures that the next time a user opens the link, it is flagged faster in the search of step 4. By adding the link to the database, the system can quickly catch the link the next time it is accessed, preventing other users from accessing the phishing link.

## 4 Algorithm Design and Implementation

The design of our algorithm is a data structure based on multiple fields and encompasses multiple functions and examinations, each designed to shed light on the different dimensions of URL credibility as described in Fig. 1.

### *4.1 URL Normalization Function*

The Remove Subdomains and Subfolders and Protocol function plays a crucial role in URL preprocessing. It employs regular expressions and string manipulation techniques to normalize a given URL by eliminating the protocol prefix (e.g., "http://", "https://"), removing the "www." prefix when present, and extracting the core domain. This normalization ensures uniformity and simplifies subsequent analysis.

Similarity assessment is a crucial aspect of URL trustworthiness evaluation. The Levenshtein Percentage function leverages the Levenshtein distance algorithm to calculate the similarity percentage between two strings. The algorithm compares two strings by checking how many characters need to be modified to change the first string into the second. The resulting similarity percentage is influenced by the lengths of the input strings and the computed Levenshtein distance.

To support the identification of potentially malicious URLs and facilitate whitelist and blacklist management, our script establishes a connection to a MySQL database. The database connection details, including the server's name, username, password, and database name, are configured within the script.

Our script includes two essential functions for managing whitelisted and blacklisted URLs. The Link in Whitelist function queries the database to determine whether the provided URL is present in the whitelist, and it returns a Boolean value indicating the URL's whitelist status.

The Add To Blacklist function allows the script to add a URL to the blacklist within the database, marking it as potentially malicious.

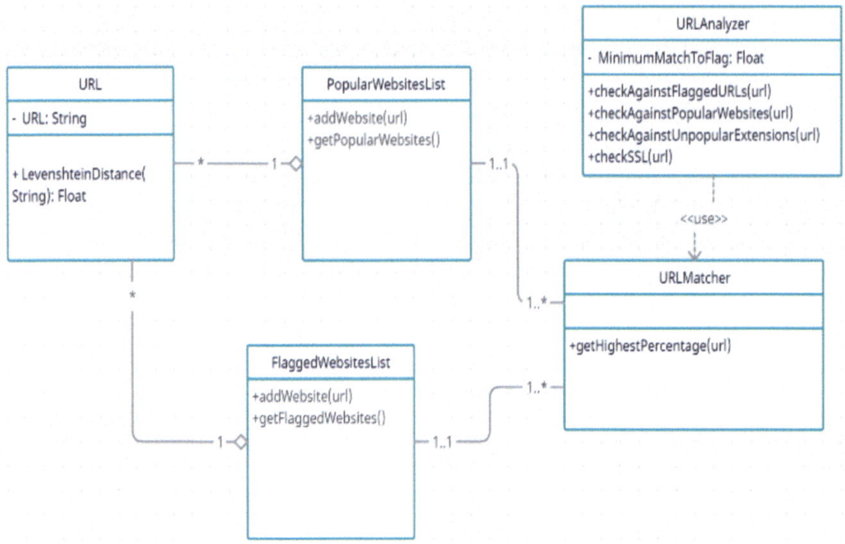


**Fig. 2** Algorithm design and implementation

The assessment of similarity percentages with respect to both blacklisted and whitelisted URLs is essential for establishing trustworthiness. Two functions, the Find Highest Similarity Percentage function and Find Highest Similarity Percentage White, have been developed: the Find Highest Similarity Percentage function retrieves URLs from the blacklist database, calculates the similarity percentages between the user-provided URL and each entry, and identifies the highest percentage among them, while Find Highest Similarity Percentage White function in analogy to its counterpart operates on the whitelist, determining the highest similarity percentage with URLs in that category (Fig. 2).

To identify potential malicious content within the URL, the Is Suspicious Text function retrieves the URL's content and performs keyword-based analysis. It scans the text for specific keywords and patterns associated with phishing, malware, fraud, adult content, counterfeit products, and other suspicious phrases. Detected keywords are compiled and reported to the user.

The hasValidSSLCertificate function utilizes cURL to ascertain the validity of the SSL certificate associated with the provided URL. It verifies the certificate against trusted authorities and returns a Boolean value, indicating its authenticity.

A trustworthy website typically provides links to its Privacy Policy and Terms of Service pages. The checkForPrivacyPolicyAndTermsLinks function retrieves the URL's content and employs regular expressions to locate these critical links. When found, the links are presented to the user, affirming the website's commitment to user data protection and terms adherence.

The age of a domain can be indicative of its trustworthiness. The getDomainAge function, leveraging WHOIS information, determines the creation date of the domain associated with the URL. It calculates and reports the domain's age in years, months, and days, contributing to the overall trustworthiness assessment.

URL redirections and popups are potential indicators of suspicious activity. The checkForRedirectionsAndPopups function employs cURL to investigate the presence of HTTP redirections and JavaScript popups on the given URL. It returns a Boolean value, indicating whether such elements were detected. Upon user input, the script initiates by normalizing the URL, ensuring it starts with "https://." Subsequently, it checks whether the URL is included in the whitelist. The user is informed that the link is whitelisted. If the URL is not whitelisted, the script proceeds to evaluate its trustworthiness. It calculates a composite trustworthiness score based on multiple factors, including similarity percentages, the presence of suspicious text, SSL certificate validity, the existence of privacy policy and terms links, and other criteria. If the calculated trustworthiness score surpasses a predefined threshold (50 in this implementation), the URL is classified as potentially malicious and is added to the blacklist.

## 5 A Comparative Analysis of the Existing Solutions and the Proposed Algorithm

In Table 1, we indicate a comparative analysis of the main solution taken into consideration for analysis in this research and our proposal. As can be seen from Table 1, our proposal provides a sustainable solution, including all the features discussed before apart from two-factor authentication, which has not been the target of our proposal because of its implementation complexity. Our implementation approach is driven by the needs of users of different ages with technical knowledge of any kind.

## 6 Results Analysis and Discussion

To test the system, we had some of the links removed from the blacklist. These are aggreg.myportfolio.com, acetax.yn, and aci.md (warning: use extreme caution when opening these links).

After removing these links, they were given to the tool to test whether or not they would be flagged as malicious by the analysis alone. As predicted, they were given a maliciousness rating within 50–60%, which was enough to flag them and add them into the database again. The details of this test are shown in Figs. 3, 4, and 5.

In Fig. 3, we use aggreg.myportfolio.com link to predict the malicious rating. The system found it highly similar to the blacklisted sites (72.73%). The system

**Table 1** Different methods identified in related works

| | [1] Data mining approach | [2] Security relevance of webpage script text | [3] Anti-phishing techniques | [4] Two-factor authentication | Proposed approach |
|---|---|---|---|---|---|
| Keyword search | Yes | Yes | Yes | No | Yes |
| User flagging | No | No | Yes | No | Yes |
| 2FA | No | No | Yes | Yes | No |
| Non-technical user friendly | No | No | Yes | No | Yes |
| Targeting chat application | Yes | No | No | No | Yes |
| Mobile-based | No | No | Yes | No | Yes |

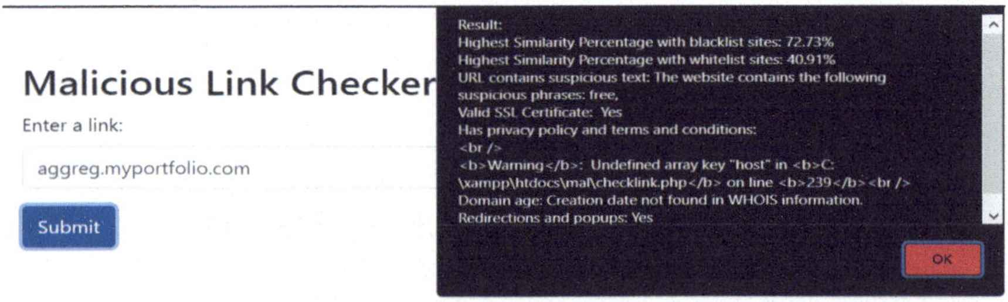


**Fig. 3** Testing case aggreg.myportfolio.com

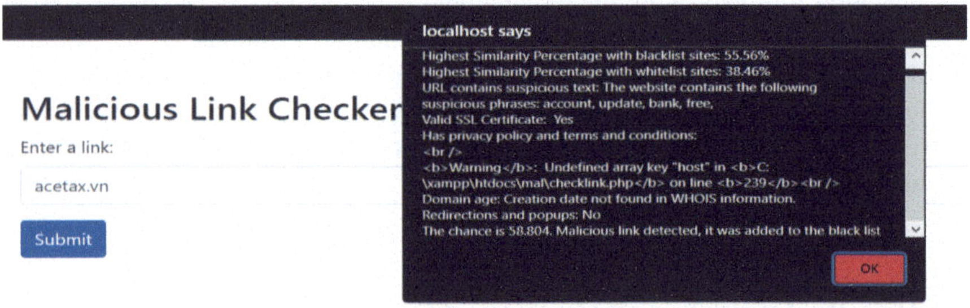


**Fig. 4** Testing case acetax.vn

detected a suspicious word (free) and did not find a terms and conditions page or privacy policy. It also detected a redirect attempt that raises the maliciousness rating. The link was deemed suspicious and added to the database.

In Fig. 4, we use acetax.yn link to predict the malicious rating. The system detected an average similarity with the blacklist (55.56%), but it found multiple suspicious words (update, bank, free). The link was deemed possibly suspicious and added to the database.

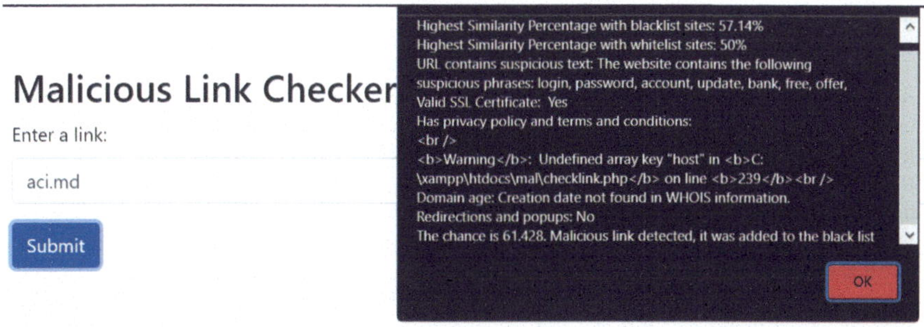


**Fig. 5** Testing case aci.md

In Fig. 5, we use aci.md link to predict the malicious rating. The system detected an average similarity with the blacklist (57.14%). However, it found many suspicious words (login, password, account, bank, free, offer). This caused it to deem the link suspicious and add it to the database.

## 7 Conclusion

Our research highlights the significance of untrustworthy URLs in messages as a serious threat to our online security in the current digital age. We acknowledge that existing solutions are either limited in scope or too complex for nontechnical users, indicating a critical need for comprehensive and user-friendly systems for URL link verification in messages.

To address this problem, we propose to design and develop an application that incorporates multiple security features to provide real-time threat detection and analysis.

Our ultimate goal for this proposed solution is to provide a secure and trustworthy system for online communication, making online experiences safer for everyone. We emphasize the importance of providing a solution that is not only effective but also user-friendly to make it accessible to a wider range of users, including those who are not technically savvy.

We believe that the system could be further enhanced with the use of machine learning. The dataset used for this system could be used as initial training data. Furthermore, a live version of the system could gather even more data for training both from user submissions and link analysis. Overall, we highlight the need for a holistic approach to address the problem of untrustworthy URLs in messages.

## References

1. Ali, M.M., Rajamani, L.: Deceptive phishing detection system: From audio and text messages in instant messengers using data mining approach. In: International Conference on Pattern Recognition, Informatics and Medical Engineering (PRIME-2012) (2012)
2. Rong, X., Jun, L., Tao, J.: Design and implementation for malicious links detection system based on security relevance of webpage script text. Int. J. Comput. Netw. Inf. Secur. (2011)
3. Sharma, P., Dash, B., Ansari, M.F.: Anti-phishing techniques – a review of cyber defense mechanisms. Int. J. Adv. Res. Comput. Commun. Eng. **11**(7) (2022)
4. Marky, K., Ragozin, K., Chernyshov, G., Matviienko, A., Schmitz, M., Mühlhäuser, M., Eghtebas, C., Kunze, K.: "nah, it's just annoying!" A deep dive into user perceptions of two-factor authentication. ACM Trans. Comput. Human Interact. (2022)
5. Nanda, A., Gupta, H.: Anti-phishing techniques in cryptography. Int. J. Electr. Comput. Eng. **5**(6), 1511 (2015). https://doi.org/10.11591/ijece.v5i6.pp1511-1515
6. Sharma, P., Dash, B., Ansari, M.F.: Anti-phishing techniques – a review of Cyber Defense Mechanisms. IJARCCE. **11**(7) (2022). https://doi.org/10.17148/ijarcce.2022.11728
7. Wu, C.-Y., Kuo, C.-C., Yang, C.-S.: A phishing detection system based on machine learning. In: 2019 International Conference on Intelligent Computing and Its Emerging Applications (ICEA) (2019)
8. Dawabsheh, A., Jazzar, M., Eleyan, A., Bejaoui, T., Popoola, S.: An enhanced phishing detection tool using Deep Learning from URL. In: 2022 International Conference on Smart Applications, Communications and Networking (SmartNets) (2022)
9. Chen, Y.Y., Yong, S., Ishak, A.: Email hoax detection system using Levenshtein distance method. J. Comput. (2014)

# Index




Ö. M. Gül et al. (eds.), *7th EAI International Conference on Robotic Sensor Networks*, EAI/Springer Innovations in Communication and Computing,
https://doi.org/10.1007/978-3-031-64495-5